AF809620

ELECTROGENESIS OF BIOPOTENTIALS

IN THE CARDIOVASCULAR SYSTEM

DEVELOPMENTS IN CARDIOVASCULAR MEDICINE

121. S. Sideman, R. Beyar and A. G. Kleber (eds.): Cardiac Electrophysiology, Circulation, and Transport. Proceedings of the 7th Henry Goldberg Workshop (Berne, Switzerland, 1990). 1991. ISBN 0-7923-1145-0.
122. D. M. Bers: Excitation-Contraction Coupling and Cardiac Contractile Force. 1991. ISBN 0-7923-1186-8.
123. A.-M. Salmasi and A. N. Nicolaides (eds.): Occult Atherosclerotic Disease. Diagnosis, Assessmentand Management. 1991. ISBN 0-7923-1188-4.
124. J. A. E. Spaan: Coronary Blood Flow. Mechanics, Distribution, and Control. 1991. ISBN 0-7923-1210-4.
125. R. W. Stout (ed.): Diabetes and Atherosclerosis. 1991. ISBN 0-7923-1310-0.
126. A. G. Herman (ed.): Antithrombotics. Pathophysiological Rationale for Pharmacological Interventions. 1991. ISBN 0-7923-1413-1.
127. N. H. J. Pijls: Maximal Myocardial Perfusion as a Measure of the Functional Significance of Coronary Arteriogram. From a Pathoanatomic to a Pathophysiologic Interpretation of the Coronary Arteriogram. 1991. ISBN 0-7923-1430-1.
128. J. H. C. Reiber and E. E. v. d. Wall (eds.): Cardiovascular Nuclear Medicine and MRI. Quantitation and Clinical Applications. 1992. ISBN 0-7923-1467-0.
129. E. Andries, P. Brugada and R. Stroobrandt (eds.): How to Face "the Faces" of Cardiac Pacing. 1992. ISBN 0-7923-1528-6.
130. M. Nagano, S. Mochizuki and N. S. Dhalla (eds.): Cardiovascular Disease in Diabetes. 1992. ISBN 0-7923-1554-5.
131. P. W. Serruys, B. H. Strauss and S. B. King III (eds.): Restenosis after Intervention with New Mechanical Devices. 1992. ISBN 0-7923-1555-3.
132. P. J. Walter (ed.): Quality of Life after Open Heart Surgery. 1992. ISBN 0-7923-1580-4.
133. E. E. van der Wall, H. Sochor, A. Righetti and M. G. Niemeyer (eds.): What is new in Cardiac Imaging? SPECT, PET and MRI. 1992. ISBN 0-7923-1615-0.
134. P. Hanrath, R. Uebis and W. Krebs (eds.); Cardiovascular Imaging by Ultrasound. 1992. ISBN 0-7923-1755-6.
135. F. H. Messerli (ed.): Cardiovascular Disease in the Elderly, 3rd ed. 1992. ISBN 0-7923-1859-5.
136. J. Hess and G. R. Sutherland (eds.); Congenital Heart Disease in Adolescents and Adults. 1992. ISBN 0-7923-1862-5.
137. J. H. C. Reiber and P. W. Serruys (eds.): Advances in Quantitative Coronary Arteriography. 1993. ISBN 0-7923-1863-3.
138. A.-M. Salmasi and A. S. Iskandrian (eds.): Cardiac Output and Regional Flow in Health and Disease. 1993. ISBN 0-7923-1911-7.
139. J. H. Kingma, N. M. van Hemel and K. I. Lie (eds.): Atrial Fibrillation, a Treatable Disease? 1992. ISBN 0-7923-2008-5.
140. B. Ostadal, N. S. Dhalla (eds.): Heart Function in Health and Disease. 1993. ISBN 0-7923-2052-2.
141. D. Noble and Y.E. Earm (eds.): *Ionic Channels and Effect of Taurine on the Heart.* Proceedings of an International Symposium (Seoul, Korea, 1992). 1993. ISBN 0-7923-2199-5.
142. H.M. Piper and C.J. Preusse (eds.): *Ischemia-reperfusion in Cardiac Surgery.* 1993. ISBN 0-7923-2241-X.
143. J. Roelandt, E.J. Gussenhoven and N. Bom (eds.): *Intravascular Ultrasound.* 1993. ISBN 0-7923-2301-7.
144. M.E. Safar and M.F. O'Rourke (eds.): *The Arterial System in Hypertension.* 1993. ISBN 0-7923-2343-2.
145. P. W. Serruys, D.P. Foley and P.J. de Feyter (eds.): *Quantitative Coronary Angiography in Clinical Practice.* 1993. ISBN 0-7923-2368-8.
146. J. Candell-Riera and D. Ortega-Alcalde (eds.): *Nuclear Cardiology in Everyday Practice.* 1993. ISBN 0-7923-2374-2.
147. P. Cummins (ed.): *Growth Factors and the Cardiovascular System.* 1993. ISBN 7923-2401-3.
148. K. Przyklenk, R.A. Kloner and D.M. Yellon (eds.): *Ischemic Preconditioning: The Concept of Endogenous Cardioprotection.* 1993. ISBN 0-7923-2410-2.

ELECTROGENESIS OF BIOPOTENTIALS

IN THE CARDIOVASCULAR SYSTEM

by

Nicholas Sperelakis

SPRINGER SCIENCE+BUSINESS MEDIA, LLC

Library of Congress Cataloging-in-Publication Data

A C.I.P. Catalogue record for this book is available from the Library of Congress.

ISBN 978-0-7923-3398-2 ISBN 978-1-4615-2590-5 (eBook)
DOI 10.1007/978-1-4615-2590-5

Copyright © 1995 by Springer Science+Business Media New York
Originally published by Kluwer Academic Publishers in 1995

All rights reserved. No part of this publication may be reproduced, stored in a retrieval system or transmitted in any form or by any means, mechanical, photo-copying, recording, or otherwise, without the prior written permission of the publisher, Springer Science+Business Media, LLC.

Printed on acid-free paper.

DEDICATION

This book is dedicated to my loyal friend and companion, Zorba-lita. Zorba is a trusting member of my family. She is my dog. Zorba-lita has been faithful through some trying times, and she has a number of accomplishments to her credits, including scaring away a burglar in the middle of the night. Zorba wants to share this dedication with the canine pets of my two daughters, Yasou (with Christina) and Calypso (with Sophia), and with the canine pet (Sambo) of my secretary, Rhonda. This most wonderful group of intelligent and heroic animals, who possess and share many of the same emotions as humans, have suffered and endured much at the hands of some members of the human race. Zorba, Yasou, Calypso, and Sambo and I fervently hope that such abuse will end forever.

ACKNOWLEDGEMENTS

The author thanks Rhonda Hentz and Humberto Ortiz-Zuazaga for typing the manuscript, Dolores Sperelakis for helping to proof the manuscript, and Anthony Sperelakis and Glenn Doerman for help with the figures. I would also like to thank Dr. Anatolii Kabakov for proofing the equations.

FOREWORD

In 1979, Dr. Sperelakis published the "Origin of the Cardiac Resting Potential", in the *Handbook of Physiology on Heart*. Since that time, many investigators and teachers in membrane biophysics have used this article as a reference source for fundamental principles and equations describing the factors which establish the resting potential in excitable and non-excitable cells. Nick has expanded the scope of this article in the present book, creating a comprehensive work and valuable reference on electrophysiological concepts underlying cellular excitability. There has long been a need for a text which precisely defines the assumptions underlying the derivations of equations which describe the principles of electrical excitability and maintenance of ion gradients in excitable cells. In this book, Nick not only defines the equations and underlying concepts of membrane potentials, but gives working examples of solutions, thus allowing investigators to utilize the fundamental principles in their research and students of membrane physiology to establish a sound basis of electrophysiological theory.

I have used the *Origin of the Cardiac Resting Potential* in graduate courses on cell physiology and biophysics, and look forward to using this new book. The time and effort required to put this work together reflects the dedication of Dr. Sperelakis to the field of membrane biophysics and electrophysiology in his long, productive career.

David R. Harder
Milwaukee

PREFACE

The idea for preparing and publishing this monograph arose because of the numerous nice comments that I have received over the years, including very recently, from graduate students, faculty, and researchers from around the world on my chapter in the 1979 Handbook of Physiology on the Heart. This book was edited by R.M. Berne and N. Sperelakis, and published by the American Physiological Society. The chapter that I wrote for that Handbook was on the origin of the resting potential in heart cells, and it was 80 pages in length (Chapter 6; pp 187-267). I had worked on writing that long chapter for almost 2 years. It contained information, explanations, equations and their derivations that were not available in any other textbook. There were also numerous tables compiling and summarizing much data from the literature. The article contained a lengthy Appendix as well that provided a review of electricity, gave derivations, and discussed some important allied topics such as the Gibbs-Donnan potentials and reversible half-cell potentials and their importance.

Because of the kind compliments over the years on that article, I decided to update and expand the scope of that article, and break it up into shorter chapters for better didactic presentation of the material. This required considerable energy and time on my part, but I felt that this effort was worthwhile in order to make this information readily available to all students. That is, the original article is "buried" in an old, expensive, and not readily-available handbook, and therefore lost or inaccessible to the vast majority of students. Only those students, whose mentors and professors knew about the article, would be referred to it. The present monograph solves this problem, as well as provides a greatly improved version of the original article. The original article has not only been updated and generalized and made more readable, but sections have also been shifted and rearranged, some sections expanded, and some new sections added. Parts of the main Appendix have been shifted to secondary appendices at the end of the appropriate chapters.

This book is intended primarily for graduate students taking courses in cell physiology, ion transport, biophysics, membrane biophysics, electrophysiology, neurophysiology, or biochemistry. These students may reside in departments of physiology, biophysics, biochemistry, neuroscience, pharmacology, biology, or zoology. In addition, advanced undergraduate students taking such courses, and medical students at the better medical schools, would find this book very helpful. Of course, faculty and researchers might find the book useful as well.

Nicholas Sperelakis
Cincinnati

TABLE OF CONTENTS

List of Appendices

List of Tables

List of Figures (1st referral)

List of Abbreviations

Na^+	sodium ion
K^+	potassium ion
Cl^-	chloride ion
Ca^{2+}	calcium ion
mM	millimolar
$[Na]_i$	intracellular Na^+ concentration (mM)
$[K]_i$	intracellular K^+ concentration (mM)
$[Cl]_i$	intracellular Cl^- concentration (mM)
$[Ca]_i$	intracellular Ca^{2+} concentration (M)
a	activity
γ	activity coefficient
a_K^i	intracellular K^+ activity
a_{Na}^i	intracellular Na^+ activity
E_{Na}	Na^+ equilibrium potential (mV)
E_K	K^+ equilibrium potential (mV)
E_{Cl}	Cl^- equilibrium potential (mV)
E_{Ca}	Ca^{2+} equilibrium potential (mV)
E_m	membrane potential
resting E_m	resting potential
RP	resting potential
E_m	true transmembrane potential immediately across membrane
$+E_{max}$	overshoot potential
AP	action potential
Å	Angstrom unit (1.0 x 10^{-8} cm)
nm	nanometers (1.0 x 10^{-9} m)
ε	dielectric constant
δ	membrane thickness (approx 70 Å)
Δx	distance across membrane (membrane thickness)
Ω	ohm
S	Siemens (Ω^{-1})

F	Faraday constant (96,500 coul/equiv.)
z	valence (equiv./mole)
zF	coul/mole
R	gas constant (8.3 joul/mole/^{o}K)
	(1.98 cal/mole/^{o}K)
T	absolute temperature (^{o}K = N° C + 273°)
μF	microfarads
μA; μAmp	microamperes
g	conductance (mho, Ω^{-1}, S)
Σg	total (summed) conductance
g_{Na}	Na^+ conductance
g_K	K^+ conductance
g_{Cl}	Cl^- conductance
g_{Ca}	Ca^{++} conductance
R	resistance (Ω)
R_K	K^+ resistance
R_{Na}	Na^+ resistance
I	current
I_{Na}	Na^+ current density (A/cm^2)
I_K	K^+ current density (A/cm^2)
I_{Cl}	Cl^- current density (A/cm^2)
I_{Ca}	Ca^{2+} current density (A/cm^2)
amp; A	ampere of current (coul/sec)
sec	second of time
R_i	resistivity of cytoplasm (Ω-cm)
R_m	membrane specific resistance (Ω-cm^2)
ρ	resistivity (Ω-cm)
ρ_m	membrane resisitivity (Ω-cm)
r_m	membrane resistance normalized for length (Ω-cm)
r_i	intracellular longitudinal resistance normalized for length (Ω/cm)
R_{in} or r_{in}	input resistance
I_o	current at point of application (x = 0)

λ	length constant
τ	time constant
t	time (sec)
$t_{1/2}$	half-time (sec)
k	rate constant (first-order) (sec^{-1})
a	fiber radius
d	fiber diameter
L	length of fiber
A_x	cross-sectional area of fiber
x	distance along fiber
vol	volume
θ	velocity of propagation
V	voltage, volts (joul/coul)
max dV/dt	maximum rate of rise of AP
N_A	Avogadro's number (6.02×10^{23} ions/mole)
coul	coulomb of charge
q_e	charge on an electron (1.60×10^{-19} coul)
equiv	equivalent of charge
dyn	dyne of force (approx. 1.0 mg)
π	geometric circle ratio of circumference/ diameter (3.14)
Π	osmotic pressure
A	area (cm^2)
A_m	membrane area (cm^2)
C	capacitance (μF)
X_c	capacitive reactance (Ω)
X_L	inductive reactance (Ω)
L	inductance (Henry)
L_m	membrane specific inductance ($H\text{-}cm^2$)
f	frequency (cycles/sec or Hertz)
f_o	resonant frequency
ω	$2\pi f$ (angular velocity)
Z_m	membrane impedance
Q	charge (coul)

I_C	capacitive current
I_L	inductive current
emf	electromotive force (V)
p.d.	potential difference (V)
ϕ	phase angle between current and voltage
ΔG	change in Gibbs free energy (joul/mol)
$(E_m - E_i)$	electrochemical driving force for any ion, i, across the cell membrane
cal	calorie of energy (1 cal = 4.17 joul)
joul	joule of energy (1 joul = 0.240 cal)
mol	moles
C_o	outside concentration (mM)
C_i	intracellular concentration (mM)
ΔC	difference in concentration
Q_{10}	temperature coefficient (ratio of rates for one temperature to that at 10° C cooler)
U	mobility $\left(\frac{cm/sec}{V/cm}\right)$
U_c	mobility of cation
U_a	mobility of anion
β	partition coefficient for an ion between solution and edge of membrane
D	diffusion coefficient (cm^2/sec)
P	permeability coefficient (cm/sec)
P_{Na}/P_K	ratio of Na^+ permeability to K^+ permeability
$\bar{\mu}$	electrochemical potential (joul/mol)
ψ_i	intracellular potential (V)
ψ_o	outside potential (V)
$\Delta\psi$	difference between inside and outside potentials; p.d.
J	flux density ($mol/sec/cm^2$)
J'	flux (mol/sec)

J_i^{Na}	Na^+ influx (mol/sec/cm^2)
J_o^K	K^+ efflux (mol/sec/cm^2)
M	molecular weight (Daltons)
e	exponential base (2.717)
log; $\log_{10}$	logarithm to the base of 10
ln	natural logarithm (base of e or 2.717)
$I_{Na/Ca}$	Na^+/Ca^{2+} exchange current
I_{TI}	transient inward current
I/V	current/voltage relationship
E_{rev}	reversal potential
E_{epp}	equilibrium potential for endplate potential
dV/dt	first time derivative of voltage
d^2V/dt^2	second time derivative of voltage
V_p	electrogenic Na^+ pump potential
ΔV_p	increase in resting E_m caused by V_p contribution
I_p	electrogenic Na^+ pump current density (μAmp/cm^2)
R_P	pump resistance
E_{diff}	net diffusion potential produced by ion concentration gradients and permeabilities
E_{ATP}^o	ATP driving potential under standard conditions
I_{bg}	background current
I_m	membrane current density (amp/cm^2)
I_l	longitudinal (axial) current
V dot	dV/dt
V double dot	d^2V/dt^2
Ag-AgCl	silver-silver chloride electrode (reversible half-cell)

h_∞	Hodgkin-Huxley h factor at infinite time or steady state
FM	frequency modulation
ECG	electrocardiogram
EMG	electromyogram
SR	sarcoplasmic reticulum
ATPase	adenosine triphosphatase

Chapter I

MEMBRANES AND ELECTRICITY

A. Introduction and Historical Aspects

Credit for the accidental discovery that an applied voltage makes muscle contract is given to Luigi Galvani (ca. 1760). Galvani found that two pieces of different metal wires touching a nerve in the leg of a freshly skinned frog caused the muscles to contract, provided that the other ends of the metal wires were in contact. Subsequently, Waller (ca. 1887) discovered that electrical currents were generated during each heartbeat and that they could be detected at the body surface. The string galvanometer was designed by Einthoven in 1913, and this instrument allowed him to record the electrocardiogram (ECG) on a routine basis. The rapidly developing science of electrocardiography then aided the physician in detecting and diagnosing various heart diseases.

The cell membrane exerts tight control over the contractile machinery in muscle cells and secretion in secretory cells. The steps that participate in these control processes are known collectively as excitation-contraction coupling (or electromechanical coupling) and excitation-secretion coupling. Therefore, it is imperative to have an understanding of the electrical properties and behavior of the cell membrane. The first step in gaining such an understanding is to examine the origin of the resting membrane potential (resting E_m).

The dependence of the resting E_m on the K^+ concentration gradient, but not on the Na^+ gradient, was first reported by Overton in 1902. Overton also observed that cells lost their excitability in the absence of Na^+. In 1902, Bernstein hypothesized that the action potential (AP) was produced by a nonspecific increase in membrane conductance to all ions; such a conductance change should cause the membrane potential (E_m) to decrease to zero. In 1939, Cole and Curtis demonstrated that there actually was nearly a 50-fold **decrease in membrane resistance** during the nerve impulse. Hodgkin and Huxley (1939), however, then found that the E_m did not just go to zero during the nerve impulse, but actually reversed in polarity (the overshoot). Hence, the Bernstein hypothesis had to be modified. The important findings of Cole and Hodgkin were enabled by the discovery of the giant nerve axon of the squid (nearly 1 mm in diameter), because relatively large intracellular electrodes could be inserted into the axoplasm.

Hodgkin and Huxley in 1952a,b proved that the large increase in conductance during the AP in squid giant axon was due first to a large increase in Na^+ conductance (g_{Na}), followed slightly later by an increase in K^+ conductance (g_K). Thus, the **increase in conductance** previously found by Cole and Curtis contained two separate components. In their experiments, Hodgkin and Huxley used the so-called **voltage clamp technique**, developed by K.C. Cole, in which the membrane potential is held (or clamped) electrically at any predetermined level by use of a **negative feedback amplifier**. By this technique, the membrane conductance increase is prevented from exploding all-or-none; that is, the *axon membrane is "tamed"*, to use the description of K.C. Cole. From the feedback current (equal to the membrane current) necessary to clamp the membrane potential at various depolarizing and hyperpolarizing steps, the membrane conductance was calculated for each voltage step. By doing the experiments both in the presence and in the absence of Na^+ (substitution by choline$^+$ ion), the part of the total conductance due to Na^+ and the time course of the Na^+ conductance change were determined by subtracting the K^+ current from the total ionic current. [Later tetrodotoxin (TTX) was used by Narahashi et al. (1964) to specifically block the fast Na^+ conductance change; the K^+ conductance change was unaffected.] Because the membrane potentials (E_m) during the clamp step and the equilibrium potentials (E_i) are known, the conductance for each ion at each point in time is calculated by dividing the current by the electrochemical driving force ($E_m - E_i$). The internal wire electrodes reduced the internal resistance to zero, thus allowing the axon to be **space clamped**, so that all parts of the membrane undergo the same potential steps simultaneously, and so there can be no longitudinal current flow.

One of the most important findings from these voltage clamp experiments was that g_{Na} is a function only of E_m and of time. Thus, the membrane potential itself controlled g_{Na} in such a fashion that a small depolarization increased g_{Na} a small amount, which in turn allowed increased Na^+ current to enter the cell, because of the large driving force for Na^+. The increased entry of Na^+ produces more depolarization, which in turn further increases g_{Na}. Thus, a **vicious cycle** or **positive feedback** occurs, producing a **regenerative increase** in g_{Na} with depolarization. For this pioneering work, Hodgkin and Huxley shared the Nobel Prize in 1963 with J.C. Eccles (for his work on synaptic transmission).

B. External Recording and Intracellular Recording

External electrical potentials were also recorded from nerve and muscle, including hearts and isolated heart tissues (such as papillary muscles) suspended in air. The presence of a resting E_m was inferred by the use of so-called **injury potential** or **demarcation potential**. Recording the potential between a crushed region and a normal region of nerve or muscle gave the injury potential, the crushed region being negative with respect to the normal region. The magnitude of the injury potential can be demonstrated to be roughly proportional to the true resting E_m (under a given set of recording conditions), the injury potential usually giving values between 30% and 60% of the true resting E_m. A current flows between the crushed end and the normal end, the so-called **injury current**. The principle of the injury potential is that the crushing destroys the selective permeability properties of the cell membrane and lowers the membrane resistance, so that an electrode placed on this area 'sees' the inside myoplasm of the cell of the uncrushed region at a distance by means of a continuous fluid pathway of relatively low resistance.

K^+ depolarization of one end of the muscle was also used to record the resting E_m between the depolarized end and the opposite end exposed to normal Ringer's solution. The principle of this method is that elevating the K^+ concentration in the Ringer's solution to about 150 mM (substitution for Na^+) lowers the resting E_m to nearly zero and lowers the membrane resistance to some extent. A **potential difference** (PD) is recorded between the two regions. Current flows externally from the normally polarized region of the muscle to the depolarized region, and internally from the depolarized region to the normally polarized region. Use of a so-called **sucrose gap** between the K^+-depolarized area and the normally polarized area increases the longitudinal resistance and allows the recording of an even greater PD between the two ends of the muscle, because of the lesser current and hence lesser voltage drop across the membrane resistances. (The sucrose gap region is perfused with an isosmolar sucrose solution to increase the resistance of the interstitial fluid space and the surface of the muscle, so that the short-circuiting of the PD between the two ends of the muscle is minimized.) As can be seen in Figure I-1, if the external resistance is made infinite, there is no short-circuiting, and the PD recorded is maximum.

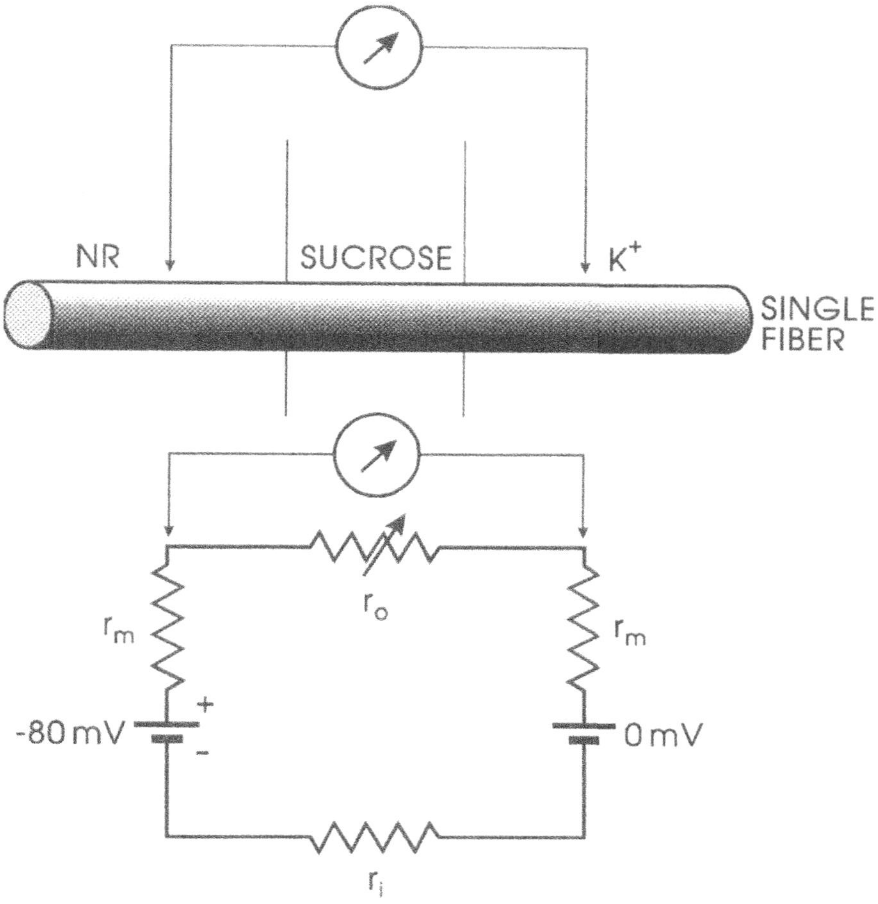

FIGURE I-1. External recording of approximate resting potential using K^+ depolarization of one end of a single fiber in a sucrose gap. Right end of fiber is bathed in a high K^+-Ringer's solution in which the NaCl is replaced by KCl to lower the resting potential in this region to zero. Voltmeter records a negative potential in the K^+-depolarized region of the fiber, with respect to the normally polarized left end. Isosmolar sucrose solution is perfused through the gap region to increase resistance of the external pathway (r_o) , and thus to minimize short-circuiting of the potential. Under usual conditions, up to about 80% of the true resting E_m can be recorded.

Electrical stimulation of the normal end allows the recording of relatively large APs.

The development of the glass micropipette electrodes by Ling and Gerard in 1949 allowed the impalement of smaller nerve and muscle cells and the recording of the true transmembrane potentials. The microelectrodes have tip diameters of about 0.2 – 0.5 μm, small enough to be inserted into single cells for intracellular recording without causing significant damage to the impaled cell. The cell membrane at the site of impalement forms a tight seal around the microelectrode tip. As illustrated in Figure I-2, if the microelectrode tip is in Ringer's solution relatively distant from the surface of a cardiac muscle preparation, no electrical potentials are recorded. The **reference (or indifferent) electrode** makes contact with the bathing solution and thereby sees the outer surface of the cell membrane. As the microelectrode (the so-called **active electrode**) is pushed up against the cell membrane where the extracellular electrical field is greater (particularly when the muscle is suspended in air), small potentials that resemble part of the standard electrocardiogram (ECG) are recorded. Upon impalement of a cell (Fig. I-2), there is a sharp drop in potential to about -80 mV, because of the difference in potential across the resting cell membrane. When the cell is resting, its interior is negative, with respect to its exterior. Transient changes in the membrane potential (i.e., the APs) are also recorded. During the rising phase of the AP, the transmembrane potential rapidly diminishes, passes through zero, and actually reverses, so that the interior of the cell becomes positive (with respect to the exterior) during the **overshoot**. The slowly-changing phase during the overshoot is the plateau phase, and this is followed by the more rapidly repolarizing phase that terminates the AP. The rising phase of the ventricular AP causes the QRS complex

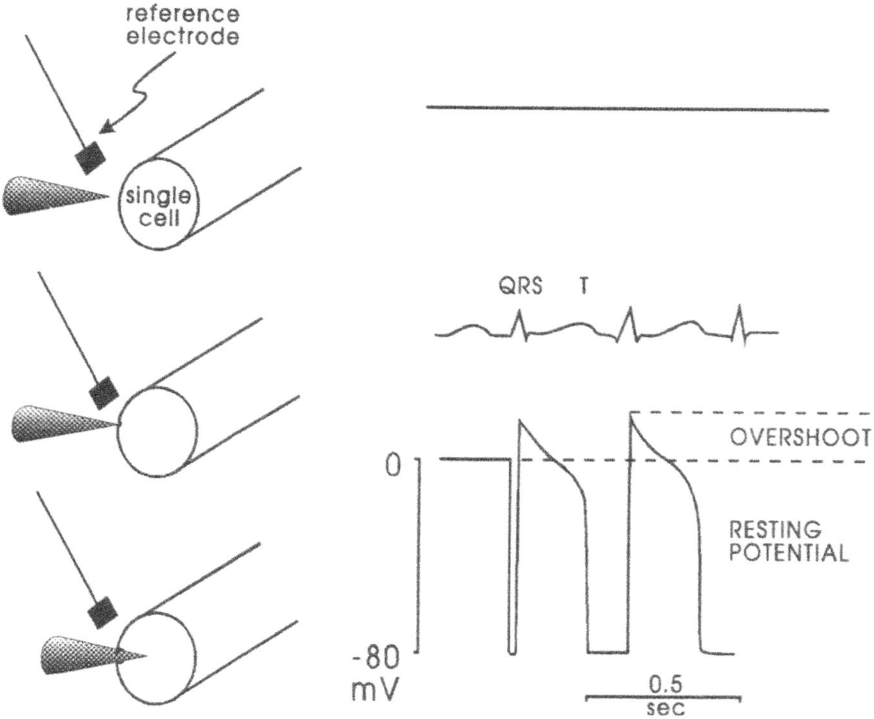

FIGURE I-2. Approach and penetration of glass capillary microelectrode into a single myocardial cell. When the microelectrode is distant from the cell, no potentials are recorded (upper row). As the microelectrode approaches very close to the cell membrane, small potentials that resemble the QRS complex and T wave of the standard electrocardiogram (ECG) are recorded (middle row). When the microelectrode penetrates into the cell, a large steady negative potential of about -80 mV (the resting potential) is recorded (lower row). Transient changes in membrane potential, the action potentials (APs) are also recorded. At the peak of the AP, the membrane potential is reversed, so that the inside of the cell, with respect to the outside, is now positive. Rapid upstroke of the AP is coincident with the QRS complex of the ECG, and the rapid repolarizing phase of the AP is coincident with the T wave.

of the ECG, and the rapid repolarizing phase produces the T wave (Fig. I-2).

Some of the electrical characteristics of nerve and muscle cells are summarized in Table I-1. The resting potentials and APs vary from one region of the heart to another. The resting potentials are about –80 to 90 mV in atrial, Purkinje, and ventricular cells, compared to –55 to 70 mV in sinoatrial (SA) and atrioventricular (AV) nodal cells. The overshoot, maximum rate of rise, and propagation velocity are also much less in nodal cells. The Purkinje fiber of the heart has the greatest rate of rise and the fastest velocity of propagation. In twitch skeletal muscle fibers, the resting E_m is about -80 mV (mammalian) or -90 mV (amphibian). In vertebrate smooth muscle cells, the resting E_m values are only about -55 mV.

C. The Cell Membrane

The resting E_m and the AP are the direct result of the special properties of the cell membrane. Davson and Danielli in 1936 proposed the theory that the cell membrane was composed of a bimolecular leaflet of phospholipid molecules sandwiched in between two layers of protein. They proposed that the nonpolar hydrophobic ends of the phospholipid molecules project toward the middle of the membrane and the polar hydrophilic ends project toward the edges of the membrane bordering on the water phases (Fig. I-3). This orientation is the thermodynamically favorable condition for the phospholipid molecules. Observations from electron microscopy indicate that the cell membrane is about 7.0–10 nm (70–100 Å)* thick, and that the phospholipid molecules found

---

* 1.0 nm = 10 Å.

TABLE I-1

Comparison of the resting potentials and action potentials of cells in different regions of the mammalian heart, skeletal muscle twitch fibers, visceral smooth muscle cells, and neurons

Parameter	Ventricular Cell	Atrial Cell	Sinoatrial Nodal Cell	Atrioventricular Nodal Cell
Resting Potential (mV)	-80 - 90	-80 - 90	-50 - 60	-60 - 70
Action Potential:				
Magnitude (mV)	110 - 120	110 - 120	60 - 70	70 - 80
Overshoot (mV)	30	30	0 - 10	5 - 15
Duration (msec)	200 - 300	100 - 300	100 - 300	100 - 300
Maximal rate of rise (V/sec)	100 - 200	100 - 200	1 - 10	5 - 15
Propagation velocity (m/sec)	0.3 - 0.4	0.3 - 0.4	<0.05	0.1
Fiber diameter (μm)	10 - 16	10 - 15	5 - 10	5 - 10

Parameter	Purkinje Fiber	Skeletal Twitch	Smooth Visceral	Neurons
Resting Potential (mV)	-90 - 95	-90	-40 - 60	-70
Action Potential:				
Magnitude (mV)	120	120 - 130	50 - 70	110
Overshoot (mV)	30	30 - 40	0 - 20	40
Duration (msec)	300 - 500	2 - 4	20 - 60	1-2
Maximal rate of rise (V/sec)	500 - 700	500 - 700	1 - 20	500-1000
Propagation velocity (m/sec)	2 - 3	3 - 5	0.03 - 0.06	1-120
Fiber diameter (μm)	100	40 - 100	3 - 5	1-20

In any species, the duration is a function of heart rate and temperature. Cell length is about 100 - 200 μm for myocardial cells, 200 - 600 μm for smooth muscle cells, and several centimeters long for skeletal muscle, and neurons can be over 1 mm in length.

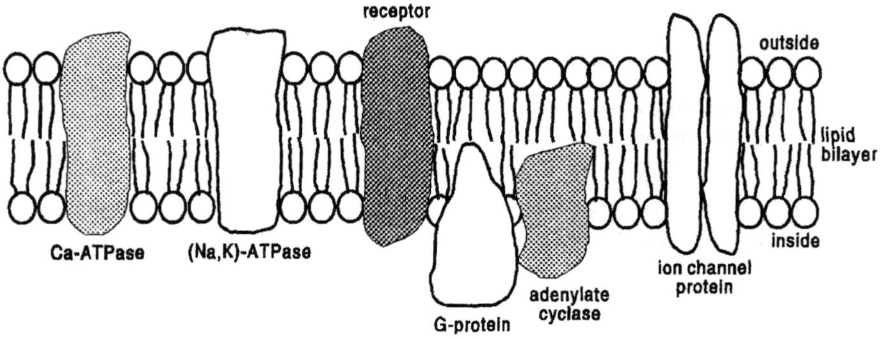

FIGURE I-3. Diagrammatic illustration of the cell membrane structure showing the lipid bilayer and some membrane proteins. Nonpolar hydrophobic tail ends of the phospholipid molecules project toward the middle of the membrane, and polar hydrophilic heads border on the water phase at each side of the membrane. Lipid bilayer is about 50-100 Å thick. For simplicity, the cholesterol molecules are not shown. Large protein molecules protrude through entire membrane thickness or are inserted into one leaflet only, as illustrated. These proteins include various enzymes associated with the cell membrane as well as membrane ionic channels. Membrane has fluidity, so that the protein and lipid molecules can move around in the plane of the membrane and fluorescent probe molecules inserted into the hydrophobic region of the membrane have freedom to rotate.

in the membranes are about the right length (3.0–4.0 nm) to stretch across half of the membrane thickness.

Studies on bilayer membranes made of various lipids and phospholipids, initiated by Mueller and Rudin (1963) and by Bangham and coworkers (1965), support the view that biological membranes have a lipid bilayer matrix. The bilayer membranes of the Mueller-Rudin type have a specific membrane capacitance (C_m) of about 0.4–1.0 $\mu F/cm^2$, which is close to the value for biological membranes. In fact, the values of C_m obtained for most biological membranes are so consistent that it is almost a universal constant (see Cole, 1968).

Calculation of membrane thickness (δ) from the following equation, assuming a capacitance (C) of 0.7 $\mu F/cm^2$ and a **dielectric constant** (ε) of 5, gives 6.3 nm (X-ray diffraction studies reveal a phosphorus-to-phosphorus spacing across the membrane thickness of about 5.2 nm; see Jain, 1972).

$$C = \frac{\varepsilon A_m}{\delta} \frac{1}{4\pi k} \qquad \text{eq. 1}$$

or

$$\delta = \frac{\varepsilon}{C/A_m} \frac{1}{4\pi k} \qquad \text{eq. 1a}$$

where A_m is the membrane area in cm^2 and k is a constant having a value of 9×10^{11} cm/F.

Most oils have dielectric constants of 3–5. The more dipolar the material, the greater the dielectric constant is. Cell membranes and bilayer membranes have similar **breakdown**

voltages of about 200 mV, equivalent to a **voltage gradient** of more than 200,000 V/cm; thus both membranes have highly effective dielectric properties. There is little doubt that the phospholipid bilayer matrix accounts for the high measured capacitance of the biological membrane. The permeability of the bilayer to water, urea, glycerol, or erythritol is within the range reported for biological membranes (see Table 3 of Thompson & Henn, 1970).

The black lipid membranes,# on the other hand, have an exceedingly high ionic resistivity (R_m between 10^6 and 10^9 $\Omega \cdot cm^2$, depending on the lipids used; see Table 2 in Thompson & Henn, 1970) that is several orders of magnitude higher than the cell membrane, i.e., about 1,000 $\Omega \cdot cm^2$. Bidirectional fluxes of ions like Cl^- may be relatively high across black lipid membranes whose membrane conductance is extremely low (Pagano & Thompson, 1968) due to exchange diffusion that would not affect membrane resistivity, which is a measure of the net movement of ions across the membrane under an electrical field.

But if the bilayer is doped with certain substances, the resistivity of the membrane can be lowered to about that of the biological membranes. For example, addition of macrocyclic antibiotics (polypeptides), known as **ionophores**, greatly lowers R_m. The ionophores may be lumped into two major classes: (a) those that act as **carriers** of ions, and (b) those that form water-filled **channels** or pores that can pass ions. For example, the valinomycin molecule is shaped like a donut whose central hole (diameter of about 0.6 nm) is just large

---

These membranes are 'black' because they are so thin that light is not reflected from their surface.

enough to accommodate a partially hydrated K^+ ion; hence valinomycin is considered to act as a so-called **cage** or carrier for K^+ through the black lipid membrane, thus lowering its resistance (raising ionic conductance). In this example, valinomycin essentially acts to increase the **lipid solubility** of K^+, because the valinomycin-K^+ complex has a high lipid solubility. The outer rim of the ring-shaped molecule is nonpolar and hence lipid soluble, whereas the inner hole surface is polar, because of six carbonyl groups ($-C = O$) that project from valine residues in the ring into the central hole. Coordination bonds are formed between the carbonyl groups and the K^+ ion.

Addition of other small proteins, such as alamethecin and excitability-inducing material (EIM), not only lowers the R_m of the bilayer, but also induces spontaneous or stimulated APs under the appropriate conditions (Mueller & Rudin, 1963). That is, this model membrane system can be made to exhibit voltage-dependent ion conductances and negative resistance characteristics typical of excitable biological membranes.

One may ask what causes the cell membrane's relatively low R_m under natural conditions. We can predict that the presence of proteins associated with the membranes must somehow account for the membranes' low resistance, compared to that of the pure phospholipid bilayers. The Davson-Danielli model of the biological membrane has been modified recently to what is generally known as the **fluid mosaic model** (Singer & Nicolson, 1972). In this model, the protein molecules protrude through the entire lipid bilayer matrix (Fig. I-3). Other protein molecules are embedded only partly into the membrane, i.e., they are interposed within the outer leaflet or the inner leaflet. The proteins in the inner leaflet may include some of the numerous enzyme activities associated with the cell membrane, such as the activity of adenylate cyclase. Some of the protein molecules protruding through the entire membrane may include the molecules associated with the membrane ion channels of resting and excited membrane, and with the (Na-K)-ATPase and Ca-ATPase molecules participating in the Na^+-K^+ linked ion pumping or Ca^{2+} pumping. The membrane proteins which serve as the receptors for the various neurotransmitters, hormones, and autocoids also appear to span through the entire membrane thickness.

Other proteins also may be adsorbed to the outer and inner surfaces of the lipid bilayer, more or less as originally conceived by Davson and Danielli. In addition, the outer surface of the cell membrane is coated with strands of mucopolysaccharides that give the cell its immunochemical properties.

The proteins that protrude entirely through the cell membrane are considered **floating**, in the sense that they can move laterally in the plane of the bilayer from one area of the membrane to another. Thus, the membrane has **fluidity**, which is the reciprocal of the **microviscosity**. The fluidity can be measured by the use of fluorescent probe molecules that lodge themselves at different levels of the lipid bilayer, e.g., in the middle of the bilayer. The ability of these probe molecules to rotate gives a measure of the microviscosity.

A high cholesterol content tends to lower the fluidity of the membrane, and a high degree of unsaturation and branching of the tails of the phospholipid molecules tend to raise the fluidity; chain length also affects fluidity. High cholesterol-containing membranes, such as the animal cell membrane, are highly condensed (because of tight molecular packing), and therefore have less fluidity (Jain, 1972). The polar portion of cholesterol lodges in the hydrophilic part of the membrane, and the nonpolar part of the planar cholesterol molecule is wedged in between the fatty acid tails, thus restricting their motion and lowering fluidity. Phospholipids with unsaturated and branched-chain fatty acids cannot be packed tightly because of steric hindrance due to their greater rigidity (Jain, 1972); hence such phospholipids increase membrane fluidity.

Low temperature decreases membrane fluidity, as might be expected. Ca^{2+} and Mg^{2+} may diminish the charge repulsion between the phospholipid head groups; this allows the bilayer molecules to pack more tightly, thereby constraining the motion of the tails and reducing fluidity (Jain, 1972). The density of packing of the phospholipid tails is about 2.0–3.0 nm per chain, and the average packing of the head group is about 6.0 nm (Jain, 1972).

One factor that determines the electrical properties and transport properties of the cell membrane must be the molecular composition of the membrane. The voltage-sensitive cation channels are composed of proteins that float in a fluid phospholipid bilayer matrix, and there is evidence that the lipid matrix influences the function of membrane proteins. For example, the hydrophobic portion of local anesthetic molecules may interpose between the lipid molecules; this separates the acyl chain tails of the phospholipid molecules further, reducing the van der Waals forces of interaction between adjacent tails, and so increasing the fluidity of the membrane. Local anesthetics are known to depress the resting conductance of the membrane for K^+ and Na^+ and to depress the voltage-dependent changes in g_{Na} and g_K , and the anesthetics also depress the voltage-dependent slow Ca^{2+} channels; that is, the local anesthetics produce a nonselective depression of most conductances of the resting and excited membrane. This depression presumably could come about indirectly by the anesthetics' effect on the fluidity of the lipid matrix. At the concentration of local anesthetics that is required to completely block excitability, the estimated concentration of local anesthetic molecules in the lipid bilayer is more than 100,000/μm^2. The depression of the (Na-K)-ATPase activity by local anesthetics (Henn & Sperelakis, 1968) also could be explained by an effect on the fluidity of the lipid matrix, although a direct effect on the protein enzyme is also possible. It has been reported that phospholipid molecules do affect the activity of the (Na-K)-ATPase enzyme [see chapter II-F on *(Na-K)-ATPase Pump*]. Thus, the lipid composition of the cell membrane also affects the membrane's electrical properties and transport characteristics.

Membrane fluidity changes are known to occur in muscle development (Herman & Fernandez, 1976) and in certain disease states, such as cancer, muscular dystrophy (Duchenne type), and myotonic dystrophy. For example, the fluidity of the membrane is reportedly increased in myotonic dystrophy (Appel & Roses, 1976). In myotonia, the skeletal muscle (twitch fiber) membrane produces repetitive spike discharges in response to a single stimulation (or a single motor end-plate potential), and consequently the muscle fiber cannot relax normally (i.e., motor nerve control over the muscle is effectively lost). The repetitive firing appears to be caused by an abnormally low Cl^- conductance of the sarcolemma (Bryant & Morales-Aguilera, 1971). Because Cl^- is off equilibrium during the AP (E_{Cl} is at the resting potential), a driving force exists for Cl^- influx during the AP. This Cl^- influx is hyperpolarizing, and acts in concert with the increase in g_K and with the turnoff of the increase in g_{Na} to bring about a sharp repolarization of the AP. In addition, one condition required for automaticity is that g_{Cl} be low. Spontaneous firing of APs and other myotonia-like symptoms are produced in isolated muscles bathed in Cl^--free solution to lower g_{Cl} (Adrian & Bryant, 1974). Animals given diazocholesterol (which interferes with cholesterol synthesis), to lower the level of cholesterol in the blood, develop myotonia-like symptoms (Sha'afi et al., 1975). In chicks with genetic muscular dystrophy, the ratio of cholesterol to phospholipid in the affected muscles is increased, and the membrane fluidity is decreased (Sha'afi et al., 1975).

Appendix I

REVIEW OF ELECTRICITY AND CABLE PARAMETERS

To understand many aspects of the electrophysiology of cardiac muscle, an appreciation of some elementary principles of electricity must be obtained. The purpose of this section is to give a brief review of the most relevant aspects of electricity.

A. Ohm's Law

The simple electrical circuit schematized in Figure I-A-1 consists of a battery, resistor, ammeter, switch, and copper wire. Closing the switch allows electrical current to flow from the positive side of the battery, through the resistor and ammeter, and back to the negative side of the battery. Electrical current is considered, by convention, as **flowing from positive to negative** through the external circuit, even though the electrons — the negatively charged particles — actually flow in the opposite direction, i.e., from negative to positive. One can consider that holes, i.e., the absence of an electron in a space that previously contained one, flow in the opposite direction of the electrons, i.e., from positive to negative around the external circuit. Within the battery, the flows are just the opposite; namely, electrons flow from positive to negative, and current flows from negative to positive.

In the circuit of Figure I-A-1, the relationship between the applied voltage, the total resistance, and the current that flows is given by **Ohm's law**. This relationship, first formulated by George S. Ohm, states that the current (I) that flows is equal to the applied voltage (V) divided by the resistance (R):

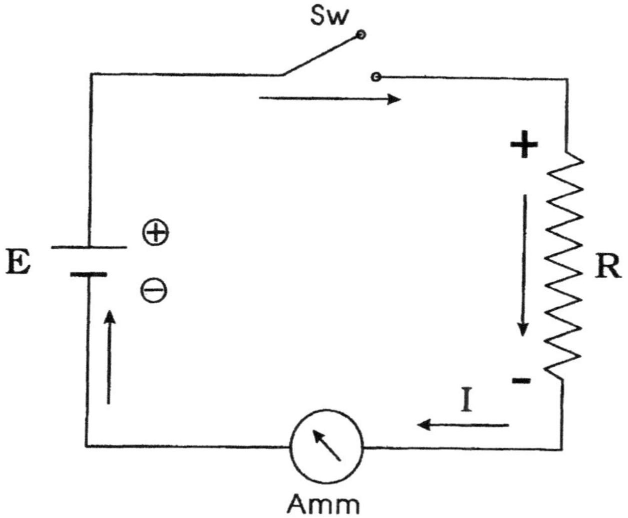

FIGURE I-A-1. A simple electrical circuit containing a battery (E) and resistance (R). When the switch (Sw) is closed, electrical current (I) flows in the direction indicated (arrows), from the positive pole of the battery to the negative pole in the external circuit, and can be measured by the ammeter (Amm) placed in series with the resistor. Current through the battery flows from the negative pole to the positive pole. Relationship between E, R, and I is given by Ohm's law: I = E/R.

$$I = \frac{V}{R} \qquad \text{eq. A-1}$$

$$\text{amperes} = \frac{\text{volts}}{\text{ohms}}$$

An **ampere** of current is the movement past a point of one coulomb of electrical charge per second (1 amp = 1 coul/sec). The unit of resistance is the **ohm**, which may be operationally defined as that value of resistance through which there is a current of 1 amp when the applied potential is 1 V ($R = V/I$). **Resistance** is a measure of the opposition to current flow; namely, for a given applied current, the greater the resistance is, the less the flow of current (inverse proportionality). Two points are at a potential difference (PD) of 1 V when it takes 1 joul of energy to transport 1 coul of charge from one point to the other (1 V = 1 joul/coul). Note that energy equals voltage times the charge (joul = V × coul). [Thus the electron-volt unit of atomic physics is a unit of energy, the charge on an electron being $.60 \times 10^{-19}$ coul.] The volt also can be operationally defined from Ohm's law as that PD necessary to produce 1 amp of current flow through a resistance of 1 Ω ($V = IR$).

A **coulomb of charge** is equal to that represented by 6.24×10^{18} electrons or univalent ions (reciprocal of $.60 \times 10^{-19}$ coul/electron). Multiplication of the charge on an electron (q_e) by **Avogadro's number** (N_A) (the number of molecules per mol) gives the number of coulombs in an equivalent (eq) of charge or mole of univalent ions

$$F = q_e \times N_A$$

$$\frac{\text{coul}}{\text{eq}} = \left(1.60\times10^{-19}\frac{\text{coul}}{\text{univalent ion}}\right)\left(6.02\times10^{23}\frac{\text{ions}}{\text{mol}}\right) \qquad \text{eq. A-2}$$

$$\cong 96{,}500\ \frac{\text{coul}}{\text{eq}}$$

This derived value is known as the **Faraday constant (F)**, used in electrochemistry for equations such as the Nernst equation.

Conductance (G) is the reciprocal of the resistance: $G = 1/R$. The greater the resistance, the lower the conductance, and vice versa. Therefore, Ohm's law can also be written as

$$I = \frac{V}{R} = \frac{1}{R}V = GV \qquad \text{eq. A-3}$$

This states that the current is a product of conductance and voltage. This form is useful for electrophysiology, since conductances in parallel simply add. The unit of conductance is the **mho** (ohm spelled backwards) or Ω^{-1} or **siemens** (S).

B. Resistors in Series

For resistors in series (Fig. I-A-2A), the total resistance (R_T) is a simple sum of the values of the resistances in series

$$R_T = R_1 + R_2 + R_3 + \cdots \qquad \text{eq. A-4}$$

If all the resistors are of the same value ($R_1 = R_2 = R_3$), then the total resistance is

$$R_T = R_1 N \qquad \text{eq. A-5}$$

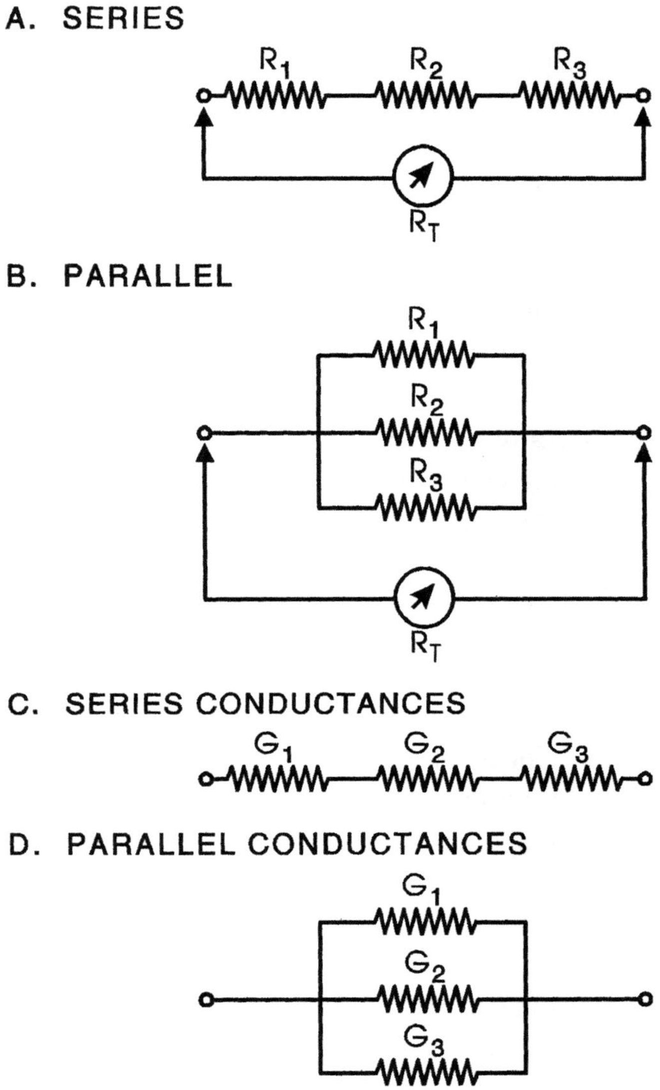

FIGURE I-A-2. A: Resistances placed in series; the total resistance (R_T) is the sum of the resistances (e.g., R_1): $R_T = R_1 + R_2 + R_3$. **B:** Resistances placed in parallel; the reciprocal of the total resistance is equal to the sum of the reciprocals of the individual resistances: $1/R_T = 1/R_1 + 1/R_2 + 1/R_3$. **C:** Reciprocal of the total conductance ($1/G_T$) of conductances in series is equal to the sum of the reciprocals of the individual conductances (e.g., G_1): $1/G_T = 1/G_1 + 1/G_2 + 1/G_3$. **D:** Total conductance of conductances in parallel is equal to the sum of the conductances: $G_T = G_1 + G_2 + G_3$.

where N is the number of resistors in series. Because myelin consists of many wrappings of membrane around the axon membrane (membranes in series), it should be obvious why myelination effectively raises membrane resistance.

C. Resistors in Parallel

For resistors in parallel (Fig. I-A-2B), the reciprocal of the total resistance ($/R_T$) is equal to the sum of the reciprocal of each resistance

$$\frac{1}{R_T} = \frac{1}{R_1} + \frac{1}{R_2} + \frac{1}{R_3} + \ldots \qquad \text{eq. A-6}$$

If all the resistors are the same value ($R_1 = R_2 = R_3 = \ldots$), then the total resistance is equal to any one resistance divided by the total number of resistors in parallel

$$R_T = \frac{R_1}{N} \qquad \text{eq. A-7}$$

If there are only two resistances in parallel, then Equation A-6 can be reduced algebraically to

$$\frac{1}{R_T} = \frac{1}{R_1} + \frac{1}{R_2} = \frac{R_1 + R_2}{R_1 R_2} \qquad \text{eq. A-8}$$

Taking the reciprocal gives

$$R_T = \frac{(R_1 R_2)}{(R_1 + R_2)} \qquad \text{eq. A-9}$$

Equation A-9 states that the total resistance of two resistors in parallel is equal to the product of the resistances divided by the sum (often called the rule of product over the sum). Note that R_T is always less than the lowest R.

D. Conductances in Series and Parallel

Conductances in series (Fig. I-A-2C) add like resistances in parallel; namely, the reciprocal of the total conductance ($1/G$) is equal to the sum of the reciprocals of each conductance (g)

$$\frac{1}{G} = \frac{1}{g_1} + \frac{1}{g_2} + \frac{1}{g_3} + \ldots \qquad \text{eq. A-10}$$

Conductances in parallel (Fig. I-A-2D) add like resistances in series; namely, the total conductance (G) is simply equal to the sum of each conductance

$$G = g_1 + g_2 + g_3 + \cdots \qquad \text{eq. A-11}$$

That this is true is demonstrated by substituting $G = 1/R$ into this equation and obtaining

$$\frac{1}{R_T} = \frac{1}{R_1} + \frac{1}{R_2} + \frac{1}{R_3} + \ldots \qquad \text{eq. A-12}$$

which is identical to the equation for resistances in parallel. It is convenient to use conductances in the study of the electrical properties of membranes, since the conductance pathways for the various species of ions are in parallel to each other and can simply be added to give the total conductance.

E. Kirchhoff's Laws

Kirchhoff's two laws may be stated in the following manner. (a) *The current going toward a point is equal to that leaving it.* For example, in a simple circuit composed of three resistors in series and a battery in series (Fig. I-A-3A), the current is constant throughout the circuit; that is, an ammeter placed anywhere in that circuit gives the same reading. In a simple parallel circuit composed of a network of three resistors in parallel connected to a series battery (Fig. I-A-3B), the current is not constant throughout. The sum of the current in each branch is equal to that in the main branch; in other words, the sum of the currents converging toward a branch point is equal to that leaving it:

$$I_T = i_1 + i_2 + i_3 \qquad \text{eq. A-13}$$

(b) *The algebraic sum of the voltage drops (IR) across each of the resistors in any closed network is equal to the source* **electromotive force** (emf), e.g., battery. For the series circuit shown in Figure I-A-3C:

$$\text{source emf} = IR_1 + IR_2 + IR_3 \qquad \text{eq. A-14}$$

The IR or voltage drop across each resistor is always in the polarity depicted; namely, the proximal side with respect to the direction of the current flow is always positive, and the distal side is negative. Current is always considered to *flow from positive to negative* in the external circuit. By convention, a positive potential is higher (or greater) than a negative one; hence, the potential is progressively dropped or lowered around the external circuit.

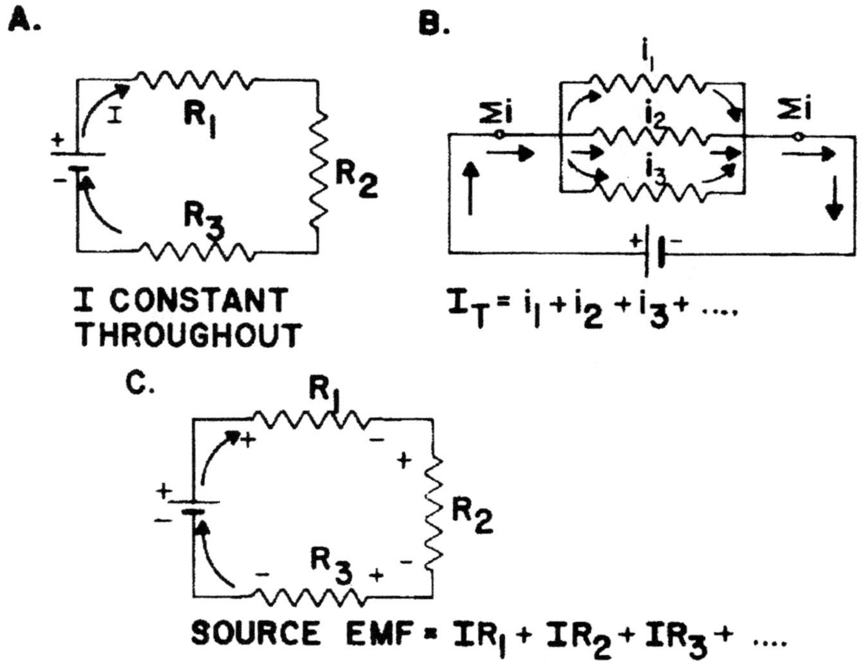

FIGURE I-A-3. Diagrammatic representation of Kirchhoff's two laws: (1) Current approaching a point is equal to that leaving the point. Thus in the simple series circuit (A) the current is constant throughout, and in the simple parallel circuit (B), the sum of the currents in each leg (e.g., 1) is equal to the total current (I_T): $I_T = i_1 + i_2 + i_3$). (2) The sum of the voltage drops (I_R) in the external circuit must equal the source emf. Thus in the simple series circuit (C), the sum of the IR drops is equal to the source emf (battery voltage): source emf = $IR_1 + IR_2 + IR_3$.

F. Nature of Capacitors

A capacitor is a device that can *store electric charge* or electricity. It consists of two parallel plates (conductors) separated by a **dielectric**. All effective dielectrics have a high electrical resistance, but the reverse is not necessarily true; namely, all high resistors are not necessarily effective dielectrics (e.g., air). The **dielectric constant** (ε) of a material is a relative index of its ability to function as an efficient dielectric compared to a vacuum, which has an ε value of 1.000. The dielectric constants of some selected material are given in Table I-A-1. The greater the **dipolar nature** of the molecules, the greater the dielectric constant. For example, water is an efficient dielectric because of its highly dipolar nature. The dipoles must be free to **flip-flop** and orient with changes in the electric field, much like the water dipole orients around small charged ions (e.g., Na^+, K^+) to form the **hydration layers**.

It is seen from **Coulomb's law**, for the force (F, in dyn) of interaction between point charges, that a large dielectric constant reduces the force of interaction between electrical charges, i.e., the **dielectric shields** or buffers

$$F = \frac{q_1 q_2}{\varepsilon d^2} \qquad \text{eq. A-15}$$

where q_1 and q_2 are the magnitudes of the charges (in statcoulomb), d is the distance between the charges, and ε is the dielectric constant. The distance squared is in Coulomb's law because of the geometrical factor: the surface area of a sphere equals $4\pi \mathbf{a}^2$ or πd^2, in which **a** is the radius and d is the diameter. Therefore, the surface area over which the total force is distributed increases with the square of the radius or

TABLE I-A-1

Dielectric constants of some materials.

Material	ε
Vacuum	1.00000
Air	1.0006
Oils	~3
Water (distilled)	80
Cell membrane	3-6
Solid insulators	2-10
Paraffin wax	2-3
Quartz	4-5
Mica	6-7
Glass	5-7
Thin layers of oxidized metal	High

diameter of the sphere. Thus, a **statcoulomb** may be defined as that charge which, when placed 1 cm from an equal and like charge in a vacuum, repels the charge with a force of 1 dyn. A **dyne** is the force necessary to give a mass of 1 g an acceleration of 1 cm/sec/sec. Because the gravitational constant is 980 $\mathrm{cm/sec^2}$, 1 dyn is equal to approximately 1 mg of force (1 g wt/980).

The magnitude of the capacitance (*C*) is equal to

$$C = \frac{\varepsilon A}{\delta}\frac{1}{4\pi k} \qquad \text{eq. A-16}$$

in which *A* is the area of each plate (in $\mathrm{cm^2}$), δ is the distance (in cm) between the plates (i.e., thickness of the dielectric), and *k* is a constant (9.0×10^1 cm/F). The unit of capacitance is the **farad** (F), named after Michael Faraday. Thus, the greater the dielectric constant, the greater the capacitance. The greater the area, the greater the capacitance. Therefore, the total capacitance of several capacitors in parallel should be the simple sum (see section 7 below (*Capacitors in Parallel*)), because the area of the plates is effectively increased. For example, the capacitance of the tuning capacitor (dial) of a radio is varied by altering the degree of overlap between two sets of interdigitating plates: the more the overlap, the greater the capacitance, because the area (*A*) is greater; air is the dielectric in this case. (Commercial capacitors may use substances such as air, oil, paper, and mica for the dielectric; the area can be made extremely large by many wrappings of double sheets of aluminum foil separated by paper or mica.)

The shorter the distance between the plates, the greater the capacitance. Therefore, the cell membrane has a very high capacitance, because the distance between the plates is only about 3.5–7.0 nm or 35–70 Å (the nonpolar phospholipid tail length of the membrane bilayer). For example, if the measured specific capacitance of the myocardial cell membrane (C_m) is 1.0 $\mu F/cm^2$, and assuming a dielectric constant of 5, then from Equation A-16.

$$\delta = \frac{A\varepsilon}{C}\frac{1}{4\pi k} = \frac{\varepsilon}{(C/A)}\frac{1}{4\pi k} \qquad \text{eq. A-16a}$$

$$= \frac{5}{1\times10^{-6}\ \mathrm{F/cm^2}}\frac{1}{(12.56)\left(9.0\times10^{11}\ \mathrm{cm/F}\right)}$$

$$= 44.2\times10^{-8}\ \mathrm{cm}$$

$$= 44\ \text{Å}$$

$$= 4.4\,\mathrm{nm}$$

Polar phospholipids in the cell membrane should have a higher dielectric constant than less polar oils.

All capacitors have a **breakdown voltage**; i.e., at more than a critical voltage level, the dielectrical material cannot withstand the PD impressed across it. Therefore the capacitor breaks down, and electric current can now flow through the dielectric material from one plate to the other. The capacitor can no longer function as a capacitance, because the two plates are almost shorted to one another. A typical breakdown voltage of a commercial radio capacitor may be 200 V. In contrast, the capacitance of the nerve or muscle membrane has a breakdown voltage of only about 200 mV. Because the two plates of the membrane capacitor are much closer (about 50 Å or 0.5×10^{-6} cm), however, the voltage gradient necessary for breakdown is $200\ \mathrm{mV}/0.5\times10^{-6}$ cm or 400,000 V/cm. At a resting potential of -80 mV, the voltage gradient is about 160,000 V/cm. Thus, the cell membrane is indeed a very effective capacitor and dielectric.

G. Capacitors in Parallel

For capacitances in parallel (Fig. I-A-4A), the total capacitance (C_T) is a simple sum of the capacitances

$$C_T = C_1 + C_2 + C_3 + \cdots \qquad \text{eq. A-17}$$

The reason for this should be clear from Equation A-16, which demonstrates that capacitance is directly proportional to the area of the plates; hence placing *capacitors in parallel effectively increases the area of the plates*, as can be deduced from Fig. I-A-4A. If the number (N) of capacitors in parallel all have the same value ($C_1 = C_2 = C_3 = \ldots$), then the total capacitance is

$$C_T = C_1 N \qquad \text{eq. A-18}$$

Thus, capacitors in parallel add like resistors in series. Invaginations of the skeletal muscle fiber or ventricular myocardial cell surface membranes to form transverse (T) tubules should increase the total measured fiber capacitance, because the increased membrane area effectively adds capacitors in parallel.

H. Capacitors in Series

For capacitances in series (Fig. I-A-4B), the reciprocal of the total capacitance ($1/C_T$) equals the sum of the reciprocals of each capacitance

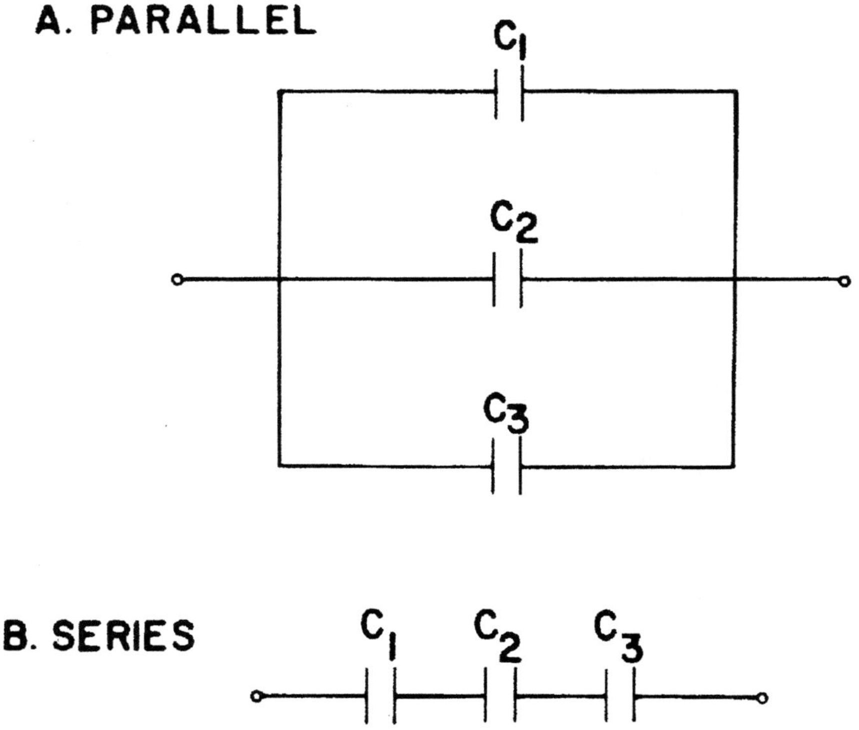

FIGURE I-A-4. A: Capacitors placed in parallel; the total capacitance is the sum of the individual capacitances (e.g., C_1): $C_T = C_1 + C_2 + C_3$. **B:** Capacitors placed in series; the reciprocal of the total capacitance is equal to the sum of the reciprocals of the individual capacitances: $1/C_T = 1/C_I + 1/C_2 + 1/C_3$.

$$\frac{1}{C_T} = \frac{1}{C_1} + \frac{1}{C_2} + \frac{1}{C_3} + \cdots \qquad \text{eq. A-19}$$

If there are several capacitors in series that are all of equal value ($C_1 = C_2 = C_3 = \cdots$), then the total capacitance is equal to any capacitance divided by N

$$C_T = \frac{C_1}{N} \qquad \text{eq. A-20}$$

If there are only two capacitors in series, regardless of values, the total capacitance equals the product over the sum:

$$C_T = \frac{C_1 C_2}{C_1 + C_2} \qquad \text{eq. A-21}$$

Note that C_T is always less than the lowest C. Because myelin consists of many (50–300) wrappings of membrane around the axon membrane, it should be obvious why myelination reduces the total capacitance of a nerve fiber. *Capacitors in series effectively increases the distance between the plates* (d), and thereby lowers the total capacitance (see eq. A-16).

I. Capacitive Reactance

Capacitors have almost infinite resistance to direct current (DC). Their **impedance** (Z, in Ω) or so-called "**AC resistance**" to sinusoidal alternating current (AC) is less, however, and is a function of the frequency (f) of the AC. The higher the frequency, the lower the capacitive reactance (X_C, in Ω). The relationship is given by

$$X_C = \frac{1}{2\pi f C} \qquad \text{eq. A-22}$$

$$= \frac{1}{\omega C}$$

in which C is capacitance and the angular velocity (ω) in radians/sec is equal to $2\pi f$. For DC, $f = 0$ and $X_C = \infty$. At ∞f, $X_C = 0$. For a given f, the larger C is, the lower X_C is.

For a muscle membrane, if C is $1\ \mu\text{F/cm}^2$ and f is 1,000 cycle/sec, then

$$X_C = \frac{1}{(6.28)(1{,}000/\text{sec})\left(1 \times 10^{-6}\ \text{F/cm}^2\right)} \qquad \text{eq. A-22a}$$

$$= 0.159 \times 10^3\ \frac{\text{sec} \cdot \text{cm}^2}{\text{F}}$$

$$= 159\ \Omega \cdot \text{cm}^2$$

J. Membrane Impedance

Because the cell membrane has a capacitive component (due to the continuous phospholipid bilayer matrix) in parallel with the resistive component (due to water-filled ion channels embedded in the lipid bilayer matrix), membrane impedance should diminish as the AC frequency is increased (Fig. I-A-5A). The relationship between frequency and relative impedance for a typical cell membrane is given in Figure I-A-5B. When f is extremely low, X_C is extremely high, and Z approaches R. As f is increased, X_C diminishes, and more and more of the current passes through C. (Current does not actually pass through the dielectric of C, but it appears so because of the phenomenon of **induced charge** that occurs.) When f is extremely high, X_C approaches 0, and therefore Z approaches 0. This is analogous to two resistors in parallel,

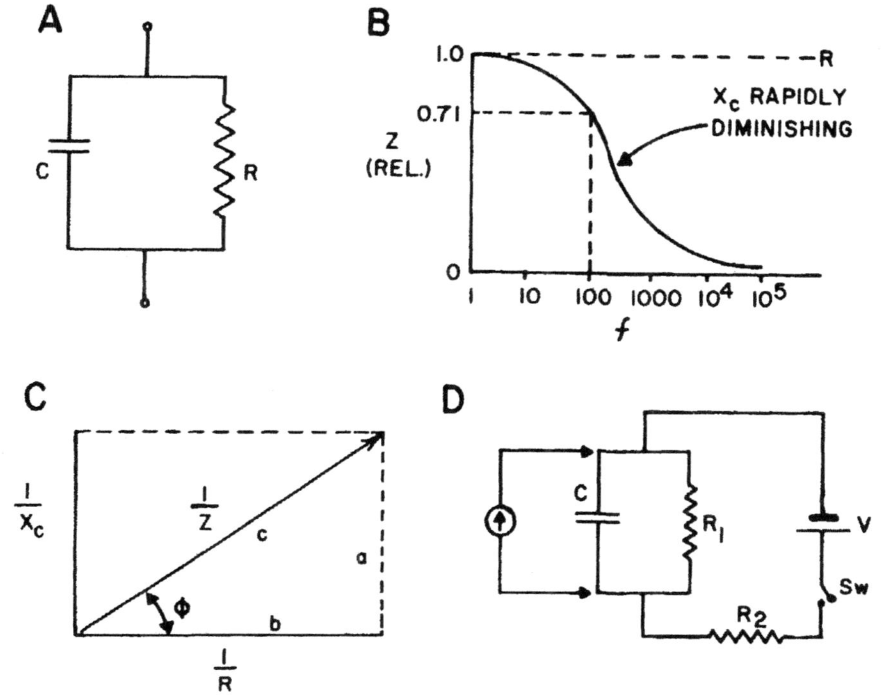

FIGURE I-A-5. A: Parallel resistance-capacitance (RC) network model of the cell membrane. **B:** Impedance (*Z*) of such a parallel resistance-capacitance network decreases as the sinusoidal frequency (*f*) is increased: a plot of *Z* vs. log frequency is an inverse sigmoidal shape. Impedance decreases because the capacitive reactance (X_C) decreases according to: $X_C = 1/2\pi fC$. **C:** Parallelogram method of obtaining the impedance of a parallel resistance-capacitance circuit. Parallelogram constructed of $1/X_C$, against *1/R* gives the vector that is equal to 1/*Z*; the phase angle (ϕ) between current and voltage is also obtained. (ϕ approaches 90° as resistance (*R*) approaches infinity; i.e., the circuit becomes purely capacitive. Also, as can be reasoned, the larger the capacitance (for a given resistance) the lower the X_C for a given *f* and hence the larger is ϕ. Thus the circuit behaves the most purely capacitive ($\phi \rightarrow 90°$), the larger the capacitance and the larger the resistance. **D:** Circuit diagram illustrates why current leads the voltage (by a certain phase angle) in a resistance-capacitance circuit. The instant the switch (Sw) is closed the frequency appears to be infinite, and so X_C is zero; therefore no voltage appears across the capacitance, although the (induced) current through the capacitance is maximum, i.e., current leads the applied voltage. R_1 and R_2, resistance 1 and 2, respectively.

one of whose impedance decreases as a function of the applied frequency. (In diathermy a very high frequency AC is used to heat damaged tissues without producing excitation of the nerve and muscle membranes; this is possible because of the capacitive component of the membrane, which causes Z_m to be extremely low and not allowing a significant membrane potential change to be produced.)

At the frequency at which $X_C = R$, however, Z is not equal to $0.5 \times R$, but is actually $0.707 \times R$, as determined by the following equation for a resistor and capacitor in parallel (note its similarity to that for two resistors in parallel, except for the square function):

$$\left(\frac{1}{Z}\right)^2 = \left(\frac{1}{X_C}\right)^2 + \left(\frac{1}{R}\right)^2 \qquad \text{eq. A-23}$$

$$Z = \sqrt{\frac{X_C^2 R^2}{X_C^2 + R^2}} \qquad \text{eq. A-23a}$$

The impedance (Z) and the phase angle (ϕ) between current and voltage can also be measured graphically by the **parallelogram method**, as illustrated in Figure I-A-5C. In this method, the *reciprocal* of the capacitive reactance is plotted on the ordinate against the *reciprocal* of the resistance on the abscissa, and the parallelogram is completed by drawing lines parallel to the two axes. The **diagonal line** is equal to the *reciprocal* of the impedance, and the angle between the diagonal and the abscissa gives the **phase angle** (ϕ) between voltage and current. In a purely resistive circuit, the current and voltage are exactly in phase ($\phi = 0°$). In a purely inductive circuit, *current lags* the voltage by 90°, whereas in a purely capacitive circuit, *current leads* the voltage by 90°. In a resistance-capacitance circuit, parallel or series, ϕ is somewhere between 0° and 90°, depending on the relative values of R and X_C.

The reason why current leads the voltage in a resistance-capacitance circuit may be discerned from Figure I-A-5D. At the instant the switch is closed, the frequency appears to be infinite (step change), and so $X_C = 0$. Therefore, there can be no potential across the capacitor (the entire voltage is dropped across R_2). Thus, the current through the capacitor is maximum, while the voltage across the capacitor is minimum, i.e., the current leads the voltage by 90°.

The identity of the parallelogram method with Equation A-23 is given by the **Pythagorean theorem** of trigonometry

$$c^2 = a^2 + b^2 \qquad \text{eq. A-24}$$

where c is the length of the hypotenuse of a right triangle, and a and b are the lengths of the other two sides. By substituting for a, b, and c from Figure I-A-5C, the following expression, which is identical to Equation A-23, is obtained

$$\left(\frac{1}{Z}\right)^2 = \left(\frac{1}{X_C}\right)^2 + \left(\frac{1}{R}\right)^2 \qquad \text{eq. A-23}$$

For a series resistance-capacitance network (Fig. I-A-6A), the corresponding analysis is

$$Z^2 = X_C^2 + R^2 \qquad \text{eq. A-25}$$

or

$$Z = \sqrt{X_C^2 + R^2} \qquad \text{eq. A-25a}$$

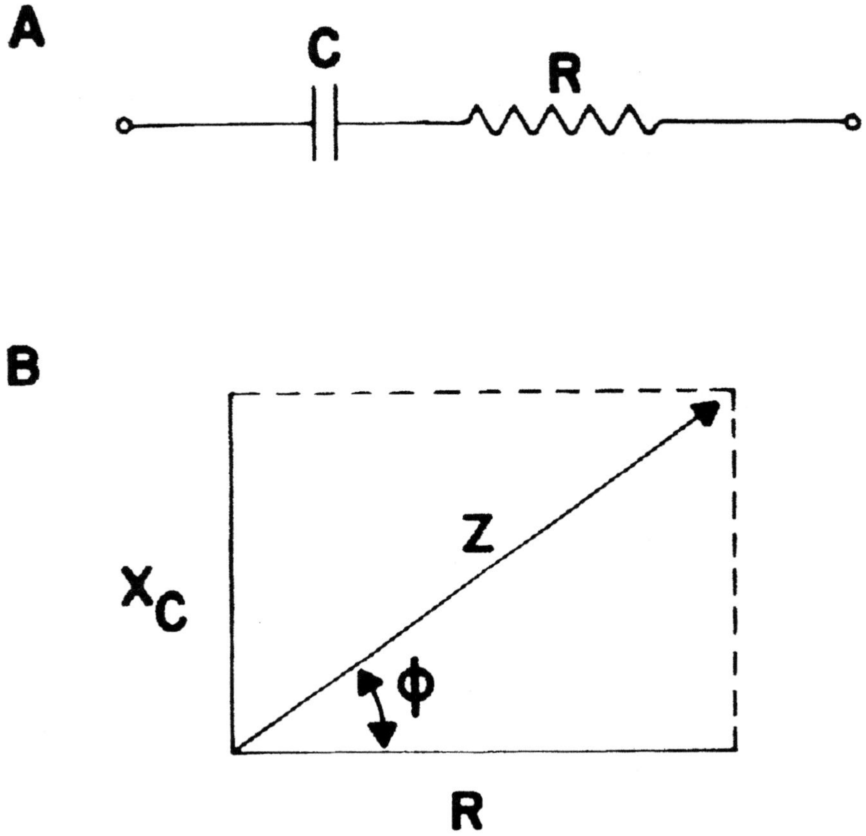

FIGURE I-A-6. A: Series resistance-capacitance (RC) network. **B:** Parallelogram method of obtaining the impedance (Z) of a series resistance-capacitance circuit. Parallelogram constructed with capacitative reactance (X_C) on the ordinite and resistance (R) on the abscissa gives the vector, which is equal to Z and the phase angle (ϕ) between current and voltage. ϕ, approaches 90° as R approaches zero or X_C approaches, infinity ($f \rightarrow 0$, or the capacitance has a lower and lower value).

which is similar to having two resistors in series, except for the square function. When $X_C = R$, then $Z = 1.414R$ (not $2 \times R$). At ∞f, $X_C = 0$, and $Z = R$. At $0f$ (DC), $X_C = \infty$, and $Z = \infty$. The parallelogram analysis is shown in Figure I-A-6B.

K. Capacitive Charge and Capacitive Current

The total charge (Q, in coul) stored on the plates of a capacitor is equal to the product of the capacitance and applied voltage

$$Q = CV \qquad \text{eq. A-26}$$
$$\text{coul} = (\text{F})(\text{V})$$

Thus, if a certain charge is placed on a capacitor (air dielectric) and then the distance between the plates is decreased so that capacitance increases, it follows that the PD across the capacitor must decrease in proportion. It also follows that because an increase in dielectric constant increases capacitance, an increase in dielectric constant must also decrease the PD for a given **charge separation**, i.e., reduces the force acting between the separated charges.

The electric charge stored in a capacitor can be used to do work. Kinetic energy is needed to charge a capacitor, the charged capacitor representing potential energy (like a battery), and discharge of the capacitor converts the potential energy back into kinetic energy. From Equation A-26 and because electric current flow (I) is a charge flowing past a point per unit of time ($I = Q/t$), the capacitive current flow (I_C, in amps) during charge or discharge of a capacitor is equal to the capacitance (in F) times the rate of change of the PD

$$I_C = C\frac{dV}{dt} \qquad \text{eq. A-27}$$

$$\frac{\text{coul}}{\text{sec}} = \text{F}\frac{\text{V}}{\text{sec}}$$

Thus for a given capacitance, the greater the rate of change of voltage, the greater the capacitive current is. In a myocardial cell membrane, for examaple, the maximum rate of change in PD across the membrane during the rising phase of the AP is relatively large (e.g., 100 mV/0.5 msec or 200 V/sec); therefore capacitive current is also relatively large. There is little or no change in membrane capacitance during an AP.

As stated above in section 10 on *Membrane Impedance*, current does not actually flow through the capacitor, but only appears to do so because of the phenomenon of **induced charge**. When electrons flow into one plate of the capacitor, the negative charges on this plate repel the electrons from the other adjacent plate, since like charges repel one another. Thus as diagrammed in Figure I-A-7, a flow of electrons into the lower plate of the capacitor leads to a flow of electrons out of the upper plate. Hence it appears as though the flow of electrons was directly through the dielectric. The deficiency of electrons in the upper plate makes the upper plate become positive, whereas the lower plate is negatively charged.

L. Membrane Time Constant

If current is passed through a simple resistance or series of resistors, the PD across the network is developed instantaneously. This is not true for a biological membrane or for the parallel resistance-capacitance network shown in Figure I-A-8A. Instead, the PD builds up in a nearly exponential (negative) manner until the final maximal value is attained, as illustrated in Figure I-A-8B. The slow charging occurs because

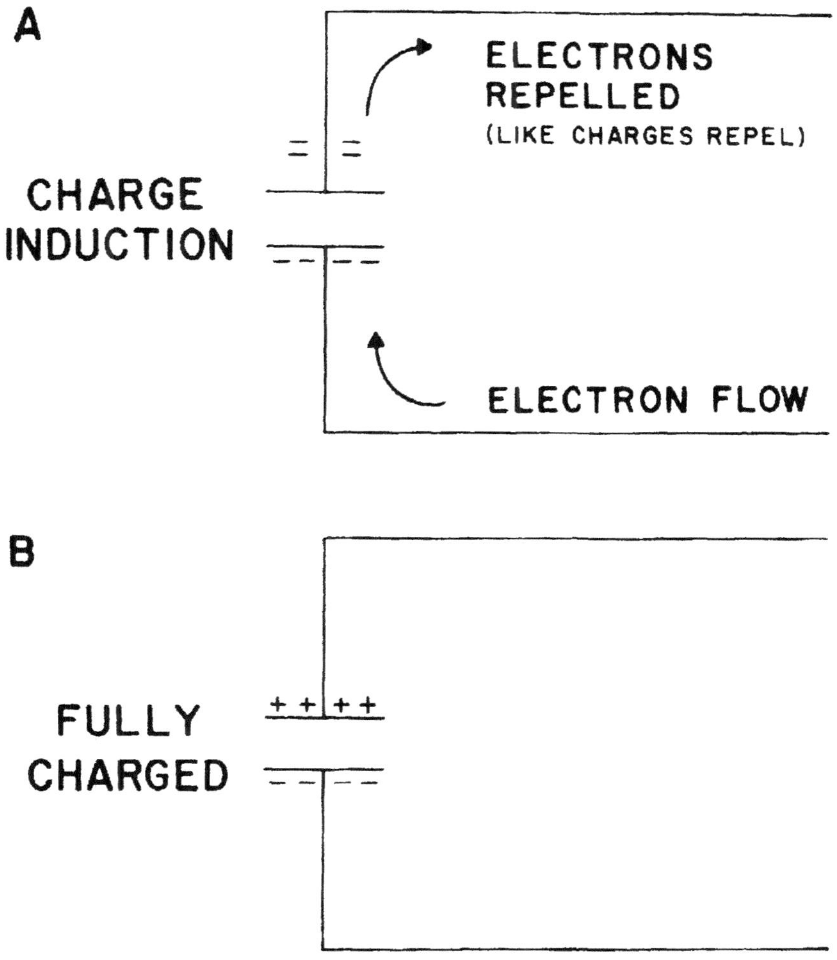

FIGURE I-A-7. Illustration of the apparent flow of current through the dielectric of a capacitor. Current does not actually flow through the capacitor but appears to because of the phenomenon of charge induction. **A:** When one plate of the capacitor is made negative, this plate repels the electrons from the other plate nearby, because like charges repel. **B:** Thus the second plate has an electron deficiency that gives the plate a positive charge.

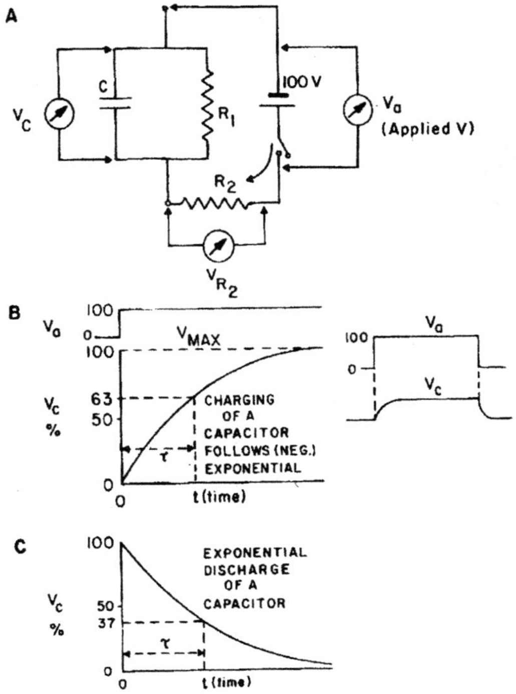

FIGURE I-A-8. A: Exponential (negative) charge and discharge of a capacitor for the Parallel resistance-capacitance (R_1C) circuit in series with a second resistance (R_2). Applied voltage is from a 100 V battery upon closing of the switch and is measured by voltmeter V_a. Voltmeter V_c measures the voltage across the capacitor, and voltmeter V_{R2} measures the voltage across R_2. **B:** Upon closing the switch, V_a suddenly jumps from 0 to 100 V. Potential recorded by V_C however, only slowly increases exponentialiy; the time constant (τ) is the time it takes for the voltage to build up to 63% ($1 - e^{-1}$) of its final (or maximal) value. The instant that the switch is closed, the frequency effectively is infinite, and so capacitive reactance (X_C) is 0; therefore no voltage appears across the parallel resistance-capacitance network, although the induced current through the capacitor is maximum. The entire voltage drop of the battery thus appears across R_2. After a long time, i.e., when the circuit reaches steady state, $f = 0$ and X is infinite; therefore the voltage across the parallel resistance-capacitance network is equal to: $(R_1/R_1 + R_2) \times 100$ V. This charging of a capacitor can also be viewed as a compartment filling with electrons and obeying first-order kinetics. **C:** Upon opening the switch, the capacitor discharges exponentially through R_1. Time constant is the time it takes for discharge to 37% (e^{-1}) of the initial (maximal) value.

of the presence of the capacitance. The time it takes for 63% of the final PD to be reached is, by definition, the **time constant** (τ). The equation describing this PD buildup is

$$V_t = V_{max}\left(1 - e^{-t/\tau}\right) \qquad \text{eq. A-28}$$

where V_t is the voltage across the network at any time (t) and V_{max} is the maximum or final PD when the steady state is reached. When $t = \tau$, then $e^{-t/\tau}$ becomes

$$e^{-1} = \frac{1}{e} = \frac{1}{2.717} = 0.37 \qquad \text{eq. A-28a}$$

Therefore,

$$\begin{aligned} V_t &= V_{max}(1 - 0.37) \\ &= 0.63V_{max} \end{aligned} \qquad \text{eq. A-29}$$

For all other values of t, Equation A-28 is solved by taking the logarithms and obtaining

$$\ln(V_{max} - V_t) = \ln V_{max} - \frac{t}{\tau} \qquad \text{eq. A-30}$$

$$V_t = V_{max} - \text{antiln}\left[\ln V_{max} - \frac{t}{\tau}\right] \qquad \text{eq. A-31}$$

$$V_t = V_{max} - \text{antilog}\frac{2.30 \log V_{max} - \dfrac{t}{\tau}}{2.30} \qquad \text{eq. A-31a}$$

Upon discharge of the resistance-capacitance network, the voltage across the capacitor (V_C) decreases exponentially, as indicated in Figure I-A-8C. Again, the time it takes for 63% of the final PD to be reached (which is to the level of 37% of the initial value) is the time constant (τ). The equation describing the decay in PD as a function of time (analogous to that for the decay in voltage as a function of distance along a cable) is

$$V_t = V_{\max} e^{-t/\tau} \qquad \text{eq. A-32}$$

where $V_{\max}$ is the maximum or initial PD at $t = 0$. When $t = \tau$, $V_t = 0.37V_{\max}$. For all other values of t, Equation A-32 is solved by taking the logarithms

$$\ln V_t = \ln V_{\max} - \frac{t}{\tau} \qquad \text{eq. A-33}$$

$$V_t = \text{antiln}\left[\ln V_{\max} - \frac{t}{\tau}\right] \qquad \text{eq. A-34}$$

$$V_t = \text{antilog}\frac{2.30 \log V_{\max} - \frac{t}{\tau}}{2.30} \qquad \text{eq. A-34a}$$

In the circuit illustrated in Figure I-A-8A, upon closing the switch, the capacitor (C) charges exponentially with a time constant about equal to R_1C (if R_1 is much greater than R_2), where R_1 and R_2 are resistance 1 and resistance 2. At the first instant after closing the switch, there is a step change in the applied potential that appears as an infinite frequency. Therefore, because $X_C = 1/2\pi fC$, $X_C \cong 0$. According to Kirchhoff's laws pertaining to the relative voltage drops across series impedances, the voltage drop is in proportion to the relative impedance (i.e., the higher impedance has the greater voltage drop), and the sum of all voltage drops must equal the source emf. Thus, all the emf of the battery should be dropped across R_2 (V_{R_2} = 100 V, where V_{R_2} is the voltage across resistor 2) and none across R_1C (V_c = 0 V). At the final steady-state value (plateau of applied voltage pulse), the applied voltage pulse is steady, and therefore the frequency is zero; hence $X_C \cong \infty$, and from analysis of the impedance of a parallel resistance-capacitance network, $Z_{R_1}C = R_1$. At this time, nearly all of the battery emf should be dropped across C (or R_1) because $R_1 >> R_2$. In this condition, the current continues to flow through R_1 and R_2. At all intermediate instants, $V_{R_2} < 100\ \text{V}$ but > 0 V, and $V_C > 0\ \text{V}$ but < 100 V. The **capacitor charges exponentially**.

One way to view why a capacitor charges and discharges exponentially is to assume that the negative plate of a capacitor is a compartment that can fill up with electrons. If so, the rate of fill-up (charging) at any moment should be a function of the actual degree of fill-up at that moment, according to the laws of **first-order kinetics**. As the concentration builds up, the rate of further buildup becomes slower and slower (a negative exponential). The process reaches equilibrium when the maximum number of electrons (maximum density or concentration of electrons) that can be stored on the plate for a given impressed voltage is reached.

Similarly, the rate of emptying of the compartment at any instant depends only on the number of electrons available at that instant. As concentration decays, the rate of further decay becomes slower and slower. At one half-time, the rate should be one-half what it was initially, because only half of the original number of electrons are now available. At the second half-time, the rate should be one-half what it was at the first half-time or one-fourth of what it was initially, and so on.

The time to build up or to decay to one-half (50%) of the final value is called the **half-time**, $t_{1/2}$. About seven half-times are necessary to reach more than 99% of the final steady-state value. The time constant is the time to build up to 63% of the final steady-state value or to decay to 37% (100% - 63%) of the initial value. About five time constants are necessary to reach more than 99% of the final value. The time constant and half-time values for a particular exponential process can be interconverted by the following relationship

$$\tau = \frac{t_{1/2}}{\ln 2} = \frac{t_{1/2}}{0.693} = 1.44\,t_{1/2} \qquad \text{eq. A-35}$$

Thus, if the membrane time constant were 1.0 msec, the half-time is 0.69 msec. The **time constant** (in seconds) and **first-order rate constant** (k, in sec^{-1}) are reciprocally related

$$k = \frac{1}{\tau} \qquad \text{eq. A-36}$$

In a nerve or skeletal muscle fiber cable (Fig. I-A-10A), the change of membrane potential during the application of electrotonic constant-current pulses is actually an **error function**, because the longitudinal resistance of the intracellular fluid is not negligible. An error function curve builds up more rapidly than does an exponential curve. The equivalent time constant for such an error function curve corresponds to a buildup to 84% of steady-state value, rather than to 63%.

M. Specific Resistance

The resistance (*R*) actually measured for any material depends on the nature of the material (its **specific resistance** or **resistivity**, ρ) and on the amount and shape of the material used. If discussion here is confined to uniform elongated cylinders or cubes of material (Fig. I-A-9A), then

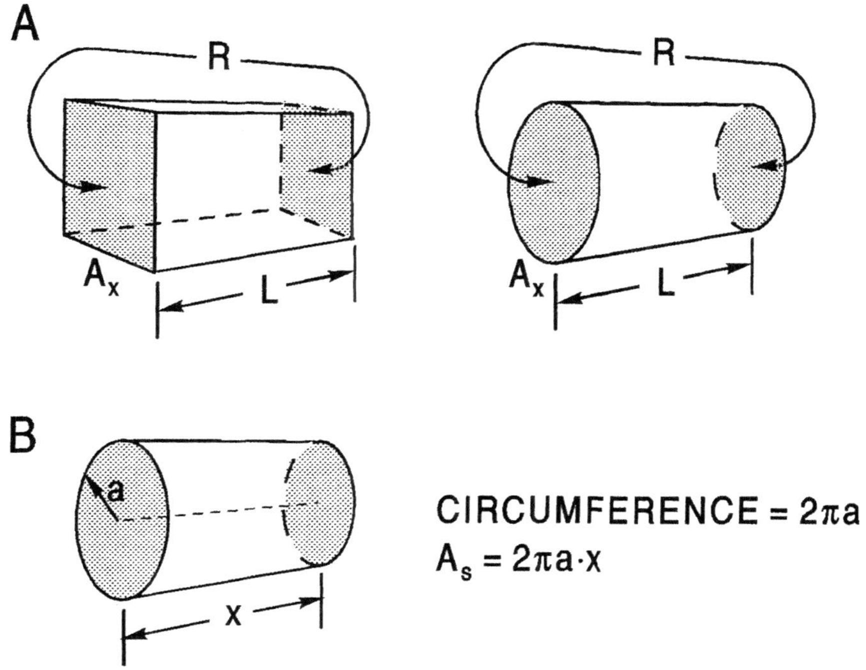

FIGURE I-A-9. Resistivity measurements on a material. **A:** For material such as myoplasm or axoplasm, the longitudinal resistance (R) depends on the resistivity of the material (R_i), the cross-sectional area (A_x or πa^2, where a is the radius) and length (L): $R = R_i \; L/\pi a^2$. Box shape and cylindrical shape are illustrated. **B:** For skin material such as cell membrane, the transverse resistance (R) depends on the so-called resistivity of the membrane R_m), and on the surface area (A_s or $2\pi a{\cdot}$x) : $R = R_m/2\pi ax$.

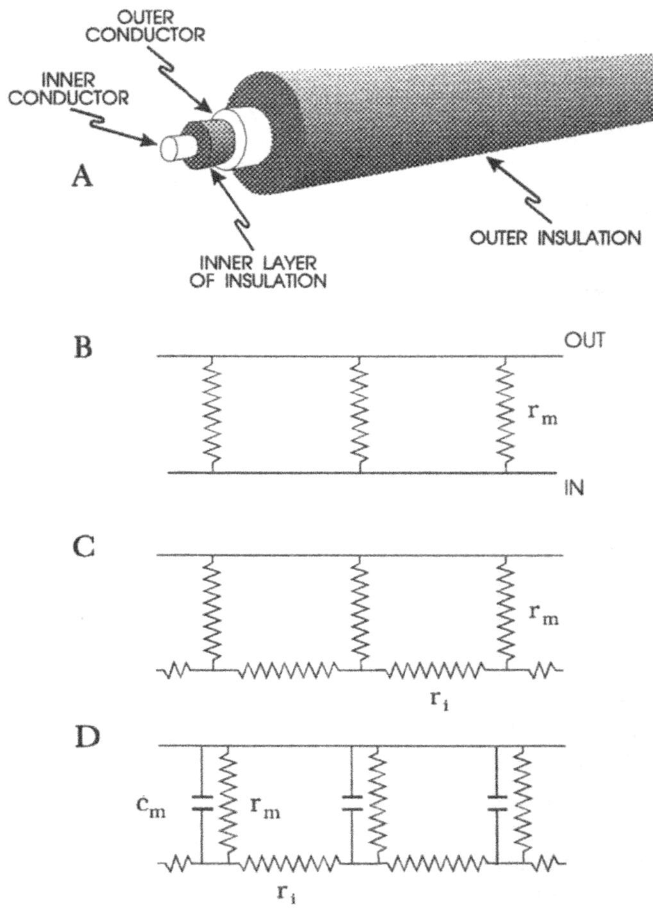

FIGURE I-A-10. Diagrams illustrating cables. **A:** Cable composed of copper wire coated with insulation and surrounded concentrically by a copper sleeve, with insulation over the latter. **B:** Simple circuit diagram for an idealized cable, which consists of two good conductors separated by a material of high resistance (*R*). **C:** For a biological cable, such as nerve fiber or skeletal muscle fiber, the inner conductor is not zero or low resistance (as would nearly be true for a copper wire), and therefore a resistance must be drawn. Even in the biological cable, however, the transverse membrane resistance (r_m is much higher than the longitudinal internal resistance (r_i)). In the biological cable, the outside resistance (r_o) can be drawn as a straight line, because r_o is relatively low in resistance compared to r_i, because the external current makes use of the entire interstitial fluid space (i.e., many resistors in parallel). **D:** Biological cable with the large membrane capacitance (C_m) included.

$$R = \rho \frac{L}{A_x} \qquad \text{eq. A-37}$$

$$\Omega = (\Omega \cdot \text{cm}) \left(\frac{\text{cm}}{\text{cm}^2} \right)$$

in which A_X is the cross-sectional area (πa^2) and L is the length. As seen from this equation, the greater the resistivity of the material, the greater is the resistance measured for any given length or cross-sectional area of material. The resistivity, by definition, is equal to the resistance a material has when the material is of unit cross-sectional area (1 cm^2) and of unit length (1 cm), hence of unit volume (1 cm^3). The units of resistivity are in Ω·cm. Rearranging Equation A-37 gives

$$\rho = R \frac{A_X}{L} \qquad \text{eq. A-37a}$$

$$\Omega \cdot \text{cm} = \Omega \frac{\text{cm}^2}{\text{cm}}$$

Mammalian Ringer solution has a resistivity of about 50 Ω·cm. The resistivity of such electrolyte solutions varies with the concentration of each of the salts present, and with the mobilities of the various species of ions (e.g., K^+, Na^+, and Ca^{2+}). The resistivity (reciprocal of conductivity or specific conductance) of solutions *decreases* with temperature elevation (whereas that of metallic conductors *increases*).

The notation is confusing because physicists use ρ to denote resistivity, whereas physiologists use R. Thus R_i is the resistivity of the internal cytoplasm (usual values of 100 – 300 Ω·cm for axoplasm and myoplasm), and R_o is the resistivity of the outside or external solution, which is normally the interstitial fluid (value of about 50 Ω·cm, comparable to that of Ringer solution). Hence, we are forced to use *r* (instead of *R*) to represent the absolute resistance (in Ω), and Equation A-37a becomes

$$R_i = r\frac{A_X}{L} \qquad \text{eq. A-38}$$

$$\Omega \cdot \text{cm} = \Omega\frac{\text{cm}^2}{\text{cm}}$$

It is convenient to substitute r_i for r/L

$$R_i = r_i A_X \qquad \text{eq. A-39}$$

$$\Omega \cdot \text{cm} = (\Omega \cdot \text{cm})\left(\text{cm}^2\right)$$

in which r_i is the internal resistance normalized for unit length, but not for unit cross-sectional area. That is, r_i is numerically equal to the internal resistance of a 1-cm length of a given cable, such as a frog sartorius fiber. Because $A_X = \pi a^2$, Equation A-39 can be modified to

$$R_i = r_i \pi a^2 \qquad \text{eq. A-39a}$$

which is the final useful form of the equation when making biological measurements (see Table I-A-2).

In dealing with the resistance of the cell surface membrane, we may proceed starting from the basic Equation A-37a and rearranging it to give

$$\rho L = R A_X \qquad \text{eq. A-37b}$$

TABLE I-A-2

Summary of the interconversions used in the measurements of properties of biological cables.

Orientation, with Respect to Fiber Axis	Actual Values Measured	Operation		Normalized for Unit Length of Fiber		Operation			Normalized for Radius of Fiber (A_x or circumference)	
Longitudinal	r (Ω)	÷ x	=	r_i	(Ω/cm)	x	πa^2	=	R_i	(Ω·cm)
Transverse	r (Ω)	× x	=	r_m	(Ω·cm)	x	$2\pi a$	=	R_m	(Ω·cm^2)
Transverse	c (F)	÷ x	=	c_m	(F/cm)	÷	$2\pi a$	=	C_m	(F/cm^2)

A_x = cross-sectional area. r = Resistance. x = Fiber length. r_i = Internal resistance per unit length of fiber. a = Fiber radius. R_i = internal resistivity. r_m = Membrane resistance for a unit length of fiber. R_m = membrane resistivity. c = Capacitance. c_m = Membrane capacitance per unit length of fiber. C_m = membrane specific capacitance.

Now because the length being considered is the thickness of the membrane (δ) and because the area being considered is the surface area (A_S) of the cylinder (Fig. I-A-9B), we must make the appropriate substitutions into Equation A-37b.

Letting $\delta = L$

$$(\rho\delta) = R\,A_X \qquad \text{eq. A-40}$$

Letting $r = R$ and $A_S = A_X$

$$(\rho\delta) = r\,A_s \qquad \text{eq. A-41}$$

Because $A_S =$ circumference ($2\pi a$) times length (x) $= 2\pi a\,x$

$$(\rho\delta) = r\,2\,\pi a\,x \qquad \text{eq. A-41a}$$

Letting $R_m = \rho\delta$

$$R_m = r\,2\,\pi a\,x \qquad \text{eq. A-42}$$

Rearranging gives

$$R_m = (rx)(2\,\pi a) \qquad \text{eq. A-42a}$$

Letting $r_m = r\,x$

$$\mathrm{R_m} = \mathrm{r_m}(2\,\pi \mathrm{a}) \qquad \text{eq. A-43}$$

$$\Omega\cdot\mathrm{cm}^2 = (\Omega\cdot\mathrm{cm})(\mathrm{cm})$$

in which r_m is the membrane resistance normalized for unit length, but not for unit surface area. That is, r_m is numerically equal to the transverse membrane resistance of a 1-cm length of a given cable (e.g., frog sartorius fiber) (Fig. I-A-10B,C). The

r_m is converted to R_m by multiplying by the circumference. These interconversions are summarized in Table I-A-2.

Although R_m is often called the membrane resistivity or specific resistance, the term is not accurate, because as just seen, the units for R_m are in Ω·cm², not Ω·cm. That is, the true resistivity has not been normalized for the unit length (thickness) of 1 cm. Thus, if R_m is 1,000 Ω·cm² and δ is 100 Å (10^{-6} cm), then the true resistivity (ρ_m) would be

$$\rho_m = \frac{R_m}{\delta} \qquad \text{eq. A-44}$$

$$= \frac{1{,}000\ \Omega\cdot\text{cm}^2}{1\times10^{-6}\ \text{cm}}$$

$$= 1\times10^{9}\ \Omega\cdot\text{cm}$$

This resistivity makes the cell membrane an effective insulator indeed, comparable to the best commercial insulators. The cell membrane, however, is extremely thin (the animal cannot afford the space to have a thick insulation (e.g., 1 cm thick) for each nerve and muscle fiber); therefore, the actual nerve cable has a great deal of energy loss (cable decrement). Note that r_m is not the true resistivity even though it has the proper units.

N. Specific Capacitance

In dealing with the capacitance of the cell surface membrane, we may proceed in a similar manner using an analysis analogous to that for R_m given above, and taking into account the fact that the total resistance of resistors added in parallel becomes smaller, whereas the total capacitance of capacitors added in parallel becomes larger. Therefore, the measured capacitance must be divided by the surface area — rather than multiplied, as for membrane resistivity — to obtain the specific capacitance. Thus, the units for specific capacitance (C_m) are $\mathrm{F/cm^2}$, compared to $\Omega\cdot\mathrm{cm}^2$ for membrane specific resistance (R_m). In addition, as with R_m, C_m is not pro-rated for 1-cm thickness, but is given for the actual thickness of the membrane; hence, the units are $\mathrm{F/cm^2}$ (and not F/cm). Thus, the equation analogous to that for membrane resistivity is parallel becomes smaller, whereas the total capacitance of capacitors added in parallel becomes larger. Therefore, the measured capacitance must be divided by the surface area — rather than multiplied, as for membrane resistivity — to obtain the specific capacitance. Thus, the units for specific capacitance (C_m) are $\mathrm{F/cm^2}$, compared to $\Omega\cdot\mathrm{cm}^2$ for membrane specific resistance (R_m). In addition, as with R_m, C_m is not pro-rated for 1-cm thickness, but is given for the actual thickness of the membrane; hence, the units are $\mathrm{F/cm^2}$ (and not F/cm). Thus, the equation analogous to that for membrane resistivity is

$$C_m = \frac{c}{A_s} = \frac{c}{2\pi a x} \qquad \text{eq. A-45}$$

where c is the absolute capacitance in F. Letting $c_m = c/x$

$$C_m = \frac{c_m}{2\pi a} \qquad \text{eq. A-46}$$

$$\frac{\mathrm{F}}{\mathrm{cm}^2} = \frac{\mathrm{F/cm}}{\mathrm{cm}}$$

where c_m (measured in F/cm) is the membrane capacitance normalized for unit length but not for unit surface area (Fig. I-A-10D). The c_m is converted to C_m ($\mathrm{F/cm^2}$) by dividing by the circumference (Table I-A-2). Thus, C_m is normalized both for unit length of the fiber and for unit circumference (radius).

O. Inductance

An inductor consists of a coil of wire with multiple turns. When current in any conductor is changed in magnitude, there is an **opposing emf induced** in any neighboring conductor, e.g., in the adjacent turns of the coil. The greater the rate of change of the current, the larger is the opposing emf induced. The opposing emf is in the opposite polarity of the applied voltage, and therefore acts to oppose the change in current. This is why an inductor is also known as a **choke**, because the inductor reacts against AC, causing a high impedance (or high so-called AC resistance) to rapidly varying currents. When the current through the coil is increasing, the induced emf opposes the current flow; whereas when the current is decreasing, the induced emf facilitates and prolongs the current. Thus, **inductance (L) tends to keep current constant** when the voltage is varied. In contrast to capacitive reactance, the **inductive reactance** (X_L , in ohms) varies directly with frequency and magnitude of the inductance

$$X_L = 2\pi f L \qquad \text{eq. A-47}$$
$$= \omega L$$

Hence, for a given inductance, the higher the frequency, the greater the reactance. The inductive reactance is zero for DC ($f = 0$). The unit of inductance is the **Henry** (H), named for the research by Joseph Henry published in 1831. In a purely inductive circuit, the current lags the voltage by a phase angle of 90°. The current through the inductance (I_L) is equal to the applied voltage (V) divided by X_L (according to Ohm's law).

The induced emf is due to the **magnetic field** that surrounds a wire that is carrying current. When current increases, the magnetic field expands, and when current decreases, the magnetic field shrinks. Thus, if the conductor is in the form of a coil, the expanding and shrinking magnetic field cuts across the neighboring turns of the wire. Whenever a moving magnetic field cuts across a conductor, an emf is generated in the conductor. (This is the principle of the **electric generator**, in which the magnetic field is usually fixed (permanent magnet) and the conductor (rotor armature) is moved.) An expanding field induces the emf in one polarity (opposing), and the shrinking field induces the emf in the opposite polarity (facilitating). For example, a large voltage and current (spark) is produced when a circuit containing an inductance is opened (broken). The phenomenon of induction of an opposing emf and a facilitating emf in a simple coil is known as **self-inductance**. The self-inductance of any given coil is a constant value, depending on the number of turns in the coil. A coil has a self-inductance of 1 H when 1 V of opposing emf is caused by a dI/dt of 1 amp/sec:

$$L = \frac{V}{(dI/dt)} \qquad \text{eq. A-48}$$

Inductance is analogous to an inertia or a mass that resists an increase or decrease in velocity. The energy (in joul) stored in the magnetic field of an inductance is equal to 0.5 LI^2 (analogous to 0.5 mass x velocity^2). **Mutual inductance** occurs between two closely spaced parallel coils, e.g., in a transformer.

The impedance for series and parallel resistance-inductance circuits are similar to those for the corresponding resistance-capacitance circuits. For a series resistance-inductance circuit

$$Z^2 = X_L^2 + R^2 \qquad \text{eq. A-49}$$

For a parallel resistance-inductance circuit

$$\left(\frac{1}{Z}\right)^2 = \left(\frac{1}{X_L}\right)^2 + \left(\frac{1}{R}\right)^2 \qquad \text{eq. A-50}$$

The same **parallelogram analysis** also applies, except by convention X_L is drawn downward (instead of upward, as in the case of X_C) to represent the fact that the current lags the voltage.

If an inductance is placed in parallel with a capacitance (parallel inductance-capacitance or so-called **tank circuit**), then the impedance follows a peculiar pattern as the AC frequency is increased. As f is increased, X_L increases, whereas X_C decreases; hence Z passes through a rather sharp maximum (approaches infinite), and the current is minimum (approaches zero), at some intermediate frequency called the **resonant frequency** (f_0). In a simple series inductance-capacitance circuit, at the resonant frequency, Z is minimum and the current is maximum. These facts are used in many electronic **tuning circuits** and **filter circuits**. The resonant frequency for either the parallel or the series inductance-capacitance circuit may be calculated from

$$f_0 = \frac{1}{2\pi}\sqrt{\frac{1}{LC}} \qquad \text{eq. A-51}$$

At low frequencies, $X_C > X_L$ and current leads voltage; at high frequencies, $X_L > X_C$ and current lags voltage. At f_0 , the current is in phase with the applied voltage and $X_C = X_L$. Thus,

$$\frac{1}{\omega C} = \omega L \qquad \text{eq. A-52}$$

or

$$\omega^2 = \frac{1}{LC} \qquad \text{eq. A-52a}$$

and

$$\omega = \sqrt{\frac{1}{LC}} \qquad \text{eq. A-52b}$$

which is the same as Equation A-51, since $\omega = 2\pi f$. If a significant resistance (in series with the inductive branch, as for the excitable membrane) is present in the parallel $(L + R)C$ circuit, then a damping factor (R^2/L^2) must be introduced, and f_0 is now given by

$$f_0 = \frac{1}{2\pi}\sqrt{\frac{1}{LC} - \frac{R^2}{L^2}} \qquad \text{eq. A-53}$$

One characteristic of the parallel inductance-capacitance tank circuit is that when suitable energy is supplied to such a circuit, **oscillations** in potential tend to occur at the resonant (or natural) frequency. In effect, at one instant, all the energy is stored in the magnetic field around the inductor; this field collapses when the current through the inductor goes to zero. This induces an emf that drives current in the opposite direction and charges up the capacitor. When fully charged, all the energy is stored as electrical charge on the plates of the capacitor. When the capacitor is fully charged and the current in the circuit goes to zero, the inductance now appears as a zero resistance ($f = 0$) across the capacitor, causing the capacitor to discharge into the inductor. This changing current induces a magnetic field in the inductor, and the cycle is repeated over and over. The oscillations produced are sinusoidal in nature. A small amount of energy needs to be put into the system; otherwise the oscillations dampen out because of the small resistance of the components and the resultant power (I^2R) loss as heat.

Oscillations of membrane potential, namely the pacemaker potentials, are known to occur in many nerve and muscle cells under natural conditions and can be made to occur in all nerve and muscle membranes under experimental conditions. The nerve membrane (e.g., squid giant axon) has been demonstrated by Cole, Mullins, and others to also contain an inductive component (see Cole, 1968). Estimates of the **apparent inductance** for biological membranes (L_m) range from 0.5–10 $\text{H}\cdot\text{cm}^2$. This inductance does not have a simple physical molecular basis, as do the resistance and capacitance components; hence it is described as an apparent inductance only. The L_m seems to be intimately associated with the K^+ resistive component of the membrane (R_K) as indicated in the circuit diagram of Figure I-A-11. As indicated by the arrow through the K^+ resistance, R_K varies as a function of the membrane potential (E_m); i.e., there is K^+ rectification. In K^+ **anomalous rectification,** when the membrane is depolarized, thereby increasing the driving force ($E_m - E_K$) for outward K^+ current, g_K decreases. Therefore, this decrease in g_K tends to keep the outward K^+ current constant, since $I_K = g_K(E_m - E_K)$, and thereby gives the behavior of an inductance.

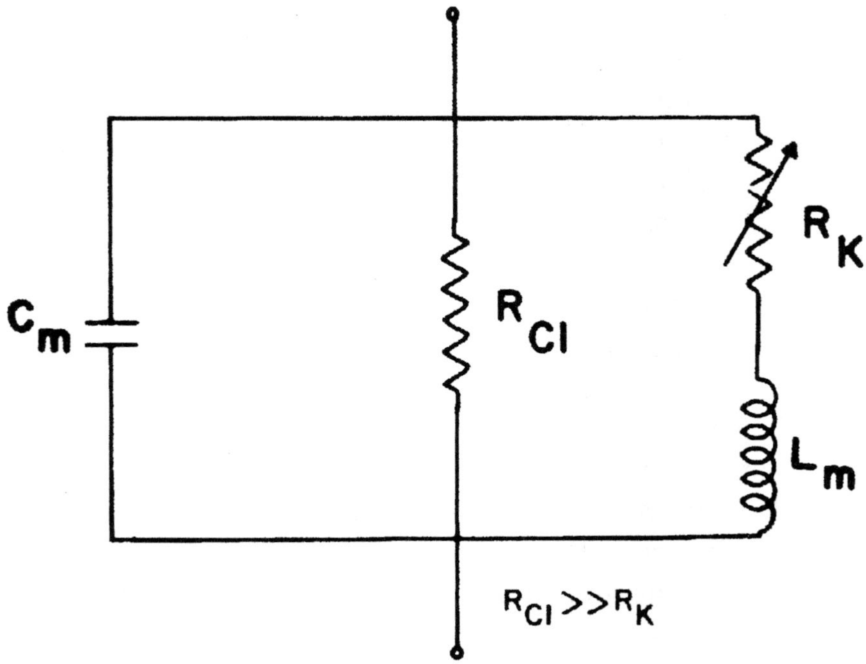

FIGURE I-A-11. Circuit diagram (K. C. Cole circuit) depicting the apparent inductance of biological membranes (L_m). L_m is in parallel to the membrane capacitance (C_m) and is due to the peculiar behavior of the K^+ resistance (R_K); that is, there is no physical basis for the inductance as there is for the capacitance (namely, the lipid bilayer matrix). When Cl^- resistance (R_{Cl}) is high compared to R_K, as is true for most myocardial cells, the parallel inductance-capacitance circuit tends to oscillate at a resonant frequency (f_o): $f_o = 1/2\pi\sqrt{1/LC}$. Arrow through R_K indicates that it is a variable resistance, depending on membrane potential. When the membrane is depolarized, R_K increases (due to the K^+ channels responsible for so-called anomalous rectification). Because $I_K = (E_m - E_K) / R_K$ the increase in driving force ($E_m - E_K$) for outward K^+ current (I_K) is counterbalanced by the increase in R_K, the latter acting to hold the K^+ current constant. This is behavior typical of an inductance and gives rise to the apparent membrane inductance.

Chapter II

ION DISTRIBUTIONS AND ION PUMPS

A. Introduction to Ion Distributions

We now focus on the ionic basis for the membrane potentials of myocardial cells. The distributions of ions across the cell membrane are illustrated in Figure II-1. The extracellular fluid or interstitial fluid has an ionic composition nearly identical to that of blood serum. The interstitial fluid has a high concentration of Na^+ (about 145 mM) and Cl^- (about 120 mM), but a low concentration of K^+ (about 4 mM) and Ca^{2+} (2 mM). In contrast, the intracellular fluid has a low concentration of Na^+ (about 15 mM or less) and Cl^- (about 4 mM), but a high concentration of K^+ (about 150 mM).

Although there has been much debate on the subject, it appears that most of the intracellular K^+ is free (that is, thermodynamically active) and has a diffusion coefficient not much less than extracellular K^+ (Kushmerick & Podolsky, 1969). In addition, the activity coefficient measured by intracellular K^+-selective glass electrodes is only slightly less than that of K^+ in free solution (Lee & Fozzard, 1975; Walker & Brown, 1977).

The intracellular free Ca^{2+} concentration in a resting cell is less than 10^{-7} M, but during contraction the concentration may rise as high as 10^{-5} to 10^{-4} M. The total intracellular Ca^{2+} (bound plus unbound) is much higher (about 2 mmol/kg), but most of this is bound to intracellular molecules, such as proteins, or sequestered into compartments, such as mitochondria and sarcoplasmic reticulum (SR).

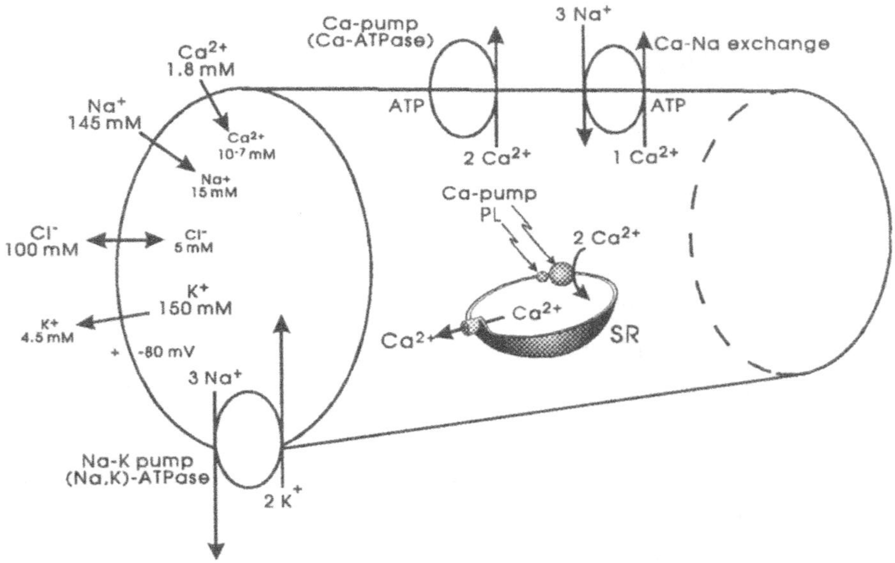

FIGURE II-1. Intracellular and extracellular ion distributions in a vertebrate myocardial cell. Also given are polarity and magnitude of the resting potential. Arrows give direction of the net electrochemical gradient. Na^{+}-K^{+} pump is located in the cell surface and presumably in the T-tubule membranes (continuations of the cell surface membrane into the cell interior). Also depicted is the Na/Ca exchanger in the sarcolemma and the two Ca pumps: one in the sarcolemma and one in the SR membrane.

Thus, under steady-state conditions the cell maintains its internal ion concentrations markedly different from those in the solution bathing the cells, as summarized in Table II-1. It is these ion concentration differences that underlie the resting potential and excitability.

B. Na^+ and K^+ Distributions

The intracellular ion concentrations are maintained different from the extracellular concentrations by the action of an active ion-transport mechanism that expends metabolic energy to push specific ions against their concentration (or electrochemical) gradients. The ion pump is located in the cell membrane at the surface of the cell, and probably also in the transverse tubular (T) membranes that are continuations of the cell membrane (Fig. II-1). The fluid in the T system is probably similar in composition to that of the bulk interstitial fluid. The major ion pump is the sodium pump, which pumps Na^+ out of the cell against the Na^+ electrochemical gradient. This pump is normally linked with K^+, however, so that while Na^+ is pumped out of the cell, K^+ is simultaneously pumped in. Thus, this pump is a Na^+-K^+ coupled pump. The K^+ is obligatory in that in K^+-free solution Na^+ can no longer be pumped out of the cell. We can see the basis for this when we discuss the properties of the (Na-K)-ATPase, the enzyme associated with the Na^+-K^+ pump.

The coupling ratio between the amount of Na^+ pumped out to the amount of K^+ pumped in theoretically may vary from $3Na^+ : 3K^+$ to $3Na^+ : 2K^+$, to $3Na^+ : 1K^+$. If the ratio is 1.0 (3 : 3), the pump would be electrically neutral or nonelectrogenic; i.e., a potential difference would not be directly produced across the membrane, because the pump would pull in three positive charges (K^+) for every three positive charges (Na^+) it pushed out. Under normal conditions, however, the pump is

TABLE II-1

Intracellular and extracellular ion concentrations in cardiac muscle.

Ion	Extracellular Concentration (mM)	Intracellular Concentration (mM)	Ratio of Extracellular to Intracellular Concentration	E_i (mV)
Na^+	145	15	9.7	+60
K^+	4	150	0.027	-94
Cl^-	120	5	24	-83
Ca^{2+}	2	10^{-7} M	2×10^4	+129

Total intracellular Ca^{2+} content is about 2 mmol/kg, but most of this is bound or sequestered in intracellular organelles (mitochondria and SR). Likewise, some of the intracellular Na^+ may be sequestered in the SR, so that the actual free Na^+ concentration in the myoplasm may be less. $[Cl^-]_i$ depends on the average membrane potential, if Cl^- is passively distributed, and therefore on heart rate.

E_i = Equilibrium potential for an ion.

electrogenic, pumping out Na^+ ions at a faster rate than K^+ ions are pumped in: $3Na^+ : 2K^+$. Thus, the pump directly produces a potential, the so-called **Na^+ pump potential**, which causes the membrane potential (E_m) to be greater (more negative) than the net diffusion potential (E_{diff}). Electrogenic pump potentials are discussed further in Chapter IX on *Electrogenic Sodium Pump Potentials*. For the present, we will ignore the electrogenic Na^+-K^+ pump.

In most excitable cells, excitability is independent from active ion transport during a short term; that is, blockade of the Na^+-K^+ pump does not immediately produce a large depolarization. [The Na^+ pump may directly or indirectly (perhaps by changing ion concentrations in restricted diffusion spaces just outside the cell membrane) influence the membrane potential for short time periods, such as during some afterpotentials in nerve.] However, during a period of minutes or hours, depending on the volume/surface area ratio of the cell, the resting potential slowly declines, because of the gradual dissipation of the ionic gradients, and eventually all excitability is lost. That is, the resting potential and excitability are ultimately dependent on the Na^+-K^+ pump. But like a water pump that pumps water into a tall water tower reservoir, breakdown of the pump does not lead to immediate cessation of water flow to the houses. The Na^+-K^+ pump can be inhibited by any procedure that lowers the intracellular ATP level, such as ischemia, hypoxia, or metabolic poisons, or that decreases energy utilization by the pump, such as inhibition of the membrane (Na-K)-ATPase with cardiac glycosides (digitalis derivatives). The pump ATPase requires Mg^{2+} for activity and is activated by Na^+ and K^+ ions, both being required. Hence, lowering the external K^+ concentration, $[K]_o$, inhibits the Na^+ pump, i.e., reduces the active Na^+ efflux. A high intracellular Na^+ concentration ($[Na]_i$) stimulates the pump.

The rate of Na^+-K^+ pumping of cells must change with the frequency of action potentials in order to maintain the intracellular ion concentrations relatively constant. For example, if the heart rate is 60 beat/min during a resting control period, and it increases to 120 beat/min during a subsequent period of exercise, then the Na^+-K^+ pump must work harder in order to maintain excitability in the new steady-state condition. In other words, if the Na^+ influx and K^+ efflux are greater, then Na^+ must be pumped back out and K^+ pumped back in at a higher rate to maintain the same ionic gradients. Therefore, the rate of pumping must be controlled. Important physiological factors that control the pump rate include the $[Na]_i$ and $[K]_o$. The pumping rate was reported to vary as a function of $[Na]_i$ raised to the first, second, or third power, depending on the type of cell. In cells that have a relatively small volume/surface area ratio, such as small-diameter nerve or muscle cells, $[Na]_i$ can increase during one or more action potential by a significant amount, compared to cells with a large ratio such as large-diameter skeletal muscle fibers. Depolarization was reported to stimulate the pump rate in some situations (Horowicz & Gerber, 1965).

C. Cl^- Distribution

In most vertebrate skeletal muscle cells studied, and perhaps in some cardiac muscles, Cl^- does not seem to be actively transported; i.e., there is no Cl^- pump. The Cl^- distributes itself **passively** (no energy used) in accordance with the membrane potential. [It has been demonstrated that Cl^- is also passively distributed in crab skeletal muscle fibers (Vaughan-Jones, 1977; Walker & Brown, 1977).] Thus $[Cl^-]_i$ is low because the negative potential inside the cell (the resting potential) pushes out the negatively charged Cl^- ion. During the action potential, however, since the inside of the cell goes in a positive direction, a greater Cl^- influx occurs because of the new electrochemical gradient for Cl^-; the magnitude of the Cl^- influx depends on the Cl^- conductance (g_{Cl}).

Thus, the exact level of $[Cl^-]_i$ in a cell that is not quiescent should depend on the frequency and duration of the action potentials, i.e., on the mean membrane potential averaged over many action potential cycles. For example, if the heart rate was 120 beats/min (with a 500 msec interval between successive action potentials) and the action potential duration was 250 msec, then E_m spends about half the time at -80 mV (the resting potential) and half the time at about +20 mV (average during the action potential plateau), thus giving an average E_m of -30 mV. Hence under these conditions, E_{Cl} should average -30 mV, rather than -80 mV, even though passively distributed, and $[Cl^-]_i$ would be about 38 mM rather than 5 mM. Thus, on the basis of finding a larger $[Cl^-]_i$ than predicted from the resting potential, one cannot conclude that Cl^- must be actively transported (inward), unless one is assured that there are no spontaneous action potentials. The E_{Cl} probably does not vary greatly during each action potential in myocardial cells (-80 mV to +20 mV) because of the low g_{Cl}.

In smooth muscle cells, however, the $[Cl]_i$ is considerably higher (e.g., 30 mM) than what would be expected from passive distribution.

D. Ca^{2+} Distribution, Exchange, and Pumping

There obviously must be some mechanism for removing Ca^{2+} from the myoplasm. Otherwise, the myocardial cell would continue to gain Ca^{2+} until there would be no electrochemical gradient for net influx of Ca^{2+}. Loss of the electrochemical gradient would not occur until the free Ca^{2+} concentration in the myoplasm was considerably greater than that outside (2 mM), because of the negative potential inside (the resting potential). Therefore, effectively there must be one or more Ca^{2+} pumps. It is well known that the SR of muscle contains a Ca^{2+}-activated ATPase (requiring Mg^{2+}, as do most ATPases) that pumps Ca^{2+} from the myoplasm into the lumen of the SR. Isolated SR from skeletal and cardiac muscles accumulates Ca^{2+} in the presence of ATP, and the reticulum is capable of pumping down the Ca^{2+} in the medium to less than 10^{-7} M. This sequestration of Ca^{2+} by the SR is essential for muscle relaxation.

Because the resting membrane is permeable to Ca^{2+}, and because an extra Ca^{2+} influx comes in with each action potential, there must be one or more ways for returning the gained Ca^{2+} back to the interstitial fluid. One major mechanism is a **Ca-pump Ca^{2+}-ATPase**, similar to that in the SR, present in the sarcolemma. Such an enzyme was recently shown to exist. However, regulation of this Ca-pump is not by phospholamban, but by another small protein.

A second major mechanism that is operative is an exchange of one internal Ca^{2+} ion for two external Na^+ ions, the **Ca^{2+}/Na^+ exchange reaction**. (In addition, it has been proposed that Ca^{2+} in the SR network is returned to the interstitial fluid across the junctional complex (or coupling) formed between the junctional SR (continuous with the network SR) and the sarcolemma (and T tubules) by some unknown means.) This reaction was discovered first in squid giant axon by Baker et al. (1969a), and is discussed in greater detail in Chapter III on *Ca^{2+}/Na^+ Exchange Reaction*. At the present, we should only note that the energy for the Ca^{2+}-Na^+ exchange reaction comes from the Na^+-K^+ pump indirectly; that is, the Na^+-K^+ pump maintains the Na^+ concentration gradient, which is used to bail out Ca^{2+}. In general, cells that gain Na^+, in some pathological or experimental condition, also gain Ca^{2+}, as expected on the basis of the exchange reaction.

A model for the Ca-ATPase of the SR is depicted in Figure II-2A. As shown, the small regulatory molecule, **phospholamban** (Gr.: "to receive phosphate"), is closely associated with the Ca-ATPase. Phospholamban has three sites (on the myoplasmic side) that can be phosphorylated by cyclic AMP-protein kinase. Phosphorylation of phospholamban results in de-repression of the enzymatic activity of the Ca-ATPase. That is, the dephosphorylated phospholamban acts as a brake on the natural activity of the Ca-ATPase. The molecular structure proposed for the SR Ca-ATPase is illustrated in Figure II-2B. As depicted, there are ten membrane-spanning domains, each containing an alpha helix structure. The first five domains have long stalks protruding into the cytoplasm. Two long cytoplasmic loops connect domains 2 and 3 and domains 4 and 5, the latter loop containing a phosphorylation region and a nucleotide binding region.

E. Energetics of Pumping

The energy cost for pumping out ions is directly proportional to the electrochemical gradient, as given by the equation

$$\Delta G_i = zF\ (E_m - E_i) \qquad \text{eq. 1}$$

where ΔG_i is the change in free energy in joules for actively transporting one mole of the ion (i) against its electrochemical driving force (E_m–E_i). The E_m is the membrane resting potential, E_i is the equilibrium potential for the ion, as calculated from the Nernst equation (see Chapter V on *Concentration Cells and Equilibrium Potentials*), z is the

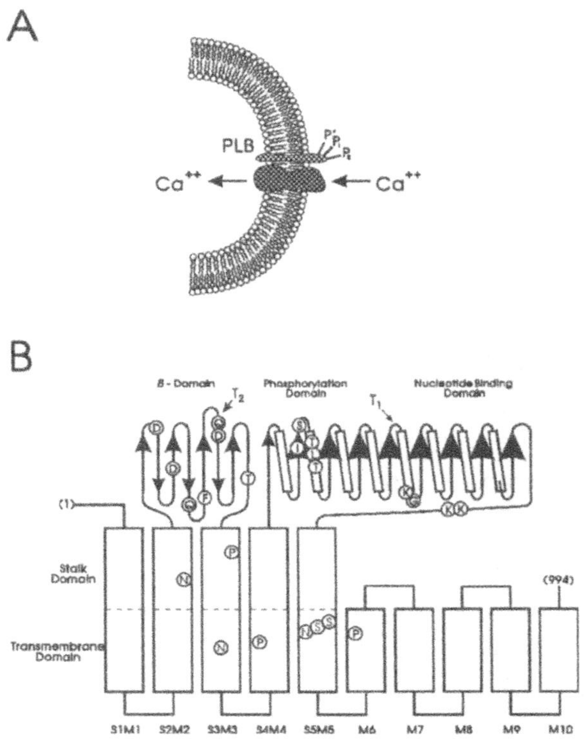

FIGURE II-2. Models for the Ca-ATPase/Ca pump of the SR,. **A:** Model depicting the relationship between phospholamban (labelled PLB) and the Ca-ATPase/Ca pump of the SR of cardiac muscle. As shown, phospholamban has three sites for phosphorylation by cAMP-PK (PK-A) on the cytoplasmic side. Phosphorylation of phospholamban stimulates the Ca pump (and sequestering of cytoplasmic Ca^{2+} into the SR lumen) by de-repression of the Ca-ATPase activity mediated by increase in the affinity of the ATPase for Ca^{2+} ion. [Figure provided courtesy of E. Kranias.] **B:** Molecular structure of the Ca-ATPase based on hydropathy analysis. There are 10 putative membrane-spanning domains (or segments) (labelled below M1 - M10), with consecutive intracellular/extracellular connective links, and a total of 994 amino acids. Each membrane-spanning domain contains an alpha-helix structure. The cytoplasmic side is represented above the domains, and the intraluminal side below the domains. The nucleotide binding region and phosphorylation region are indicated on the long cytoplasmic loop connecting between the 4th and 5th domains. As depicted, the first 5 domains have long stalk regions as extensions protruding into the cytoplasm (labelled below S1 - S5). [Reproduced in modified form from D.H. MacLennan (1990) with permission.]

valence of the ion (sign, and eq/mol), and F is the Faraday constant (96,500 coul/eq). For example, if the resting potential is -80 mV and E_{Na} is +60 mV, then the energy cost of pumping out Na^+ is

$$\Delta G_{Na} = z_{Na} F\left(E_m - E_{Na}\right) \qquad \text{eq. 1a}$$

$$\Delta G_{Na} = \left(+1\frac{\text{eq}}{\text{mol}}\right)\left(9.65\times10^4\,\frac{\text{coul}}{\text{eq}}\right)[-0.080\text{ V}-(+0.060\text{ V})]$$

$$= \left(9.65\times10^4\,\frac{\text{coul}}{\text{mol}}\right)(-0.140\text{ V})$$

$$= -13{,}500 \quad \text{joul / mol}$$

since joul = V × coul. Because 1 cal = 4.18 joul, the energy cost becomes 3,240 cal/mol. Doing the same calculation for K^+ (assume E_K = -94 mV)

$$\Delta G_K = z_K F\left(E_m - E_K\right) \qquad \text{eq. 1b}$$

$$\Delta G_K = \left(+1\frac{\text{eq}}{\text{mol}}\right)\left(9.65\times10^4\,\frac{\text{coul}}{\text{eq}}\right)[-0.080\text{ V}-(-0.094\text{ V})]$$

$$= \left(9.65\times10^4\,\frac{\text{coul}}{\text{mol}}\right)(+0.014\text{ V})$$

$$= 1{,}350 \quad \text{joul/mol}$$

$$= 324 \quad \text{cal / mol}$$

Therefore, in this example for a cell under physiological conditions, it is about 10 times more costly to pump out Na^+ ions than to pump in K^+ ions, and this difference is directly due to the electrochemical driving force for Na^+, which is 10-fold larger than that for K^+ ions. The difference in signs for the two calculations comes about because the Na^+ and K^+ driving forces are in opposite directions.

The Na^+-K^+ pump is known to pump 3 Na^+ ions per molecule of ATP hydrolyzed (regardless whether the transport is against or down the electrochemical gradient) (Sen & Post, 1964; Garrahan & Glynn, 1967a). If 3 K^+ ions are pumped concomitantly, then the total energy required per mole of ATP is: 3 × 3,240 cal/mol for Na^+ plus 3 x 324 cal/mol for K^+ equals 10,692 cal/mol. This value is close to the free energy available from the hydrolysis of 1 mol of ATP under the conditions existing inside the cell. Estimates of the standard free energy (ΔG^o) for hydrolysis of ATP to ADP + P_i (Kushmerick & Podolsky, 1969) range from -8,800 to -9,100 cal/mol (25° C, pH 7.0, 1 mM Mg^{2+}, ionic strength of 0.2); at the ATP, ADP, and P_i concentrations prevailing in the cell, the actual free energies can be substantially greater. The ΔG^o for hydrolysis of creatine phosphate is about −11,400 cal/mol.

Equation 1 can be simply derived from the general equation for the free energy change resulting from both electrical work (zFE_m) and osmotic work ($RT \ln[c^+]_i/[c^+]_o$)

$$\Delta G_i = RT \ln \frac{[c^+]_i}{[c^+]_o} + zFE_m \qquad \text{eq. 2}$$

where R is the gas constant, T is the absolute temperature, and $[c^+]_i$ and $[c^+]_o$ are the internal and outside cation concentrations. From the **Nernst equation**

$$E_i = \frac{-RT}{zF} \ln \frac{[c^+]_i}{[c^+]_o} \qquad \text{eq. 3}$$

Substituting $-zFE_i$ for $RT \ln[c^+]_i/[c^+]_o$ in Equation 2 gives

$$\Delta G_i = (-zFE_i) + zFE_m \qquad \text{eq. 4}$$

which is the same as Equation 1.

F. (Na-K)-ATPase/Pump

1. Introduction

An ATPase that was activated by Na^+ and K^+ ions was first discovered in nerve membrane in 1957 by Skou. Since then, this ATPase has been identified in virtually all tissues examined, including that of the heart. The enzyme requires both Na^+ and K^+ ions for activation, activity being zero in the absence of either cation. Because of this requirement, it is generally agreed that this enzyme intimately participates in the coupled active transport of Na^+ and K^+ across cell membranes. The (Na-K)-ATPase is sometimes called the **Na-K pump ATPase** or the **transport ATPase**. The enzyme also requires Mg^{2+} ion for activity, as do most ATPases, and (Na-K)-ATPase is inhibited by Ca^{2+} ion (and Ba^{2+} and Zn^{2+}). ATP, Mg^{2+}, and Na^+ are required at the inner surface of the membrane in the intact cell, and K^+ is required on the outer surface. There is competition between Na^+ and K^+ for the inside site at which Na^+ activates, as well as for the outside site at which K^+ activates. The Na-K affinity ratios for the inside and outside sites differ greatly. The (Na-K)-ATPase is specifically inhibited by cardiac glycosides (digitalis derivatives), such as ouabain, that act on the outer surface. This enzyme belongs to the P-type of ion-motive ATPases, in which a phosphate group is transfered from ATP to an aspartate residue in the enzyme. The Ca-ATPase and H/K-ATPase belong to this same class.

2. Enzyme preparations

The enzyme preparation generally studied by most investigators is a membrane vesicular fraction obtained by differential centrifugation (100,000 g for 1 hr) of tissue homogenates. The 100,000 g fraction mainly contains membranes arranged as spherical vesicles, whose diameters generally vary between 0.1 and 0.5 μm. The membranes include cell membranes as well as intracellular membranes, which in the case of muscle are mainly SR membranes. Separation of the cell membrane from the SR membrane in the 100,000 g vesicular preparation is difficult in muscle. But lysed red blood cells, which do not contain intracellular membranes, do make a rather homogeneous cell membrane preparation, and many studies on the (Na-K)-ATPase have been performed using this preparation. If the vesicular membrane preparation is extracted by NaI, then most of the ouabain-insensitive ATPase activity is removed. For example, in membranes taken from chick embryonic hearts, ouabain (1 mM) produced a 92% inhibition of the total ATPase activity in the NaI-extracted preparation, compared with only a 34% inhibition in the nonextracted preparation (Sperelakis & Lee, 1971).

Studies have been made on a purified (Na-K)-ATPase protein, removed from the membranes by use of detergents and obtained from tissues exceedingly rich in (Na-K)-ATPase, such as kidney (outer medulla), salt gland of birds, rectal gland of dogfish shark, and the electrical organ of Electrophorus (see 1975 review by Skou). SDS polyacrylamide gel electrophoresis of purified (Na-K)-ATPase protein from mammalian kidney gives two major protein bands (Kyte, 1974; Jorgensen, 1975). One band is sharp and has a molecular weight of about 100,000; the other band is more diffuse, contains a sialoglycoprotein, and has a molecular weight of about 57,000. In other preparations, the molecular weight of the phosphorylatable protein ranged between 84,000 and 139,000, whereas that of the glycoprotein ranged between 35,000 and 57,000 (see Table I of 1975 review by Schwartz et al.). The molar ratio of the high molecular-weight protein to the low molecular-weight protein were 1 : 1 (canine kidney), 1 : 2 (canine kidney), and 2 : 1 (dogfish rectal gland). Estimates of the molecular weight of the (Na-K)-ATPase ranged from 175,000 to 250,000, and there may be two catalytic proteins per functional molecule of (Na-K)-ATPase (Kyte, 1975; Giotta, 1976).

3. Specific activity

There are about 3.0-5.7 nmol of cardiac glycoside binding sites per mg protein. The number of Na-K pump sites per cell can be increased by loading the cells with Na^+ or by exposing them to low concentrations of ouabain for long periods. This effect is prevented by inhibitors of protein synthesis, thus suggesting that the cells can be stimulated to produce and incorporate more (Na-K)-ATPase protein into the cell membrane (Boardman et al., 1975).

The specific activity of the purified enzyme varies between about 700 and 7,000 μmol hydrolyzed ATP/mg protein per hr, depending on the source of the enzyme. In heart muscles, there is at least a 10-fold increase in specific activity in the purified preparations over vesicular preparations. The molar activity of the purified preparations varies between 1,700 and 12,000 μmol ATP/μmol protein per min. The calculated **turnover numbers** range between 1,100 and 6,500/min (18–108/sec), which is slow compared to other enzymes.

4. Lipid surround

The (Na-K)-ATPase protein is intimately associated with the membrane lipid structure in which it is embedded, and there is evidence that the lipid matrix affects the activity of the enzyme [see 1975 review by Skou]. In lipid-depleted systems the enzymatic activity is nearly abolished; acidic phospholipids have a reactivating effect, the most potent of which is phosphatidylserine. For example, break-down of phosphatidylserine leads to complete inactivation, and addition of phosphatidylserine or phosphatidic acid restores activity. Cholesterol seems to be required for enzyme activity. In general, the composition of the lipid matrix influences the conformation, and thereby the activity of the protein enzyme. The fluidity of the lipids in the membrane depends on the length of the fatty acid chains, the degree of unsaturation, and the cholesterol : phospholipid ratio, and fluidity has an effect on the temperature-dependence of the enzyme activity (Kimelberg & Papahajopoulos, 1972).

5. Temperature dependence

Arrhenius plots of (Na-K)-ATPase activity (log activity vs. 1/T(°K)) reveal a break at about 20° C [references in review by Schwartz (1962)], presumably due to a transition in the state of the lipids between a less-ordered state (higher temperatures) to a more-ordered state (lower temperatures). It was reported that phospholipase-A treatment converted the Arrhenius plot to a single straight line, and that addition of phosphatidylserine restored the break (Charnock et al., 1973). Molecules of (Na-K)-ATPase in a more-ordered lipid matrix would be less active and would require a higher activation energy (greater Q_{10}). Thus, any agent that increases the fluidity of the membrane should increase the activity of the enzyme. It has been suggested that the fluidity of membranes changes during embryonic development (Herman & Fernandez, 1976) and during aging (Barnett & Palazzotto, 1974). Therefore, the increase in specific activity of the (Na-K)-ATPase of the heart during embryological development (Sperelakis, 1972) could be due to both an increase in the number of enzyme molecules and an increase in membrane fluidity.

6. Reconstitution experiments

Incorporation of purified (Na-K)-ATPase into artificial lipid bilayer membranes has been successful. Jain et al. (1972) added enzyme to one side of a planar black lipid membrane (formed from oxidized cholesterol and alkanes), and they found changes in electrical properties of the membrane when ATP was added to the same side as the enzyme; ouabain added to the opposite side inhibited the changes. It was also reported that purified enzyme could be incorporated into lipid vesicles; ATP added to the suspending medium induced movement of Na^+ into the vesicles, and this movement was blocked by ouabain inside (but not outside) the vesicles (Hokin, 1974). The enzyme molecules may not need to all insert into the lipid bilayer in the same direction (Goldin, 1977).

7. Cytochemical localization

By cytochemical technique, Schultze and Wollenberger (1969) demonstrated the presence of a ouabain-sensitive ATPase activity in the myocardial cell membrane. Forbes & Sperelakis (1972) demonstrated that the SR membrane of cardiac muscle and skeletal muscle contains ATPase activity that is inhibited by ouabain, i.e., presumably a (Na-K)-ATPase. The reaction product PbS appears in the lumen of the SR, the equivalent of the outer surface of the cell membrane. Parts of the SR also contain a ouabain-insensitive ATPase activity, presumably a Ca^{2+}-ATPase.

8. Evidence for enzyme and pump being the same

There is strong evidence that the (Na-K)-ATPase participates in the active transport of Na^+-K^+: (a) The enzyme requires both Na^+ plus K^+ for activity. The Na-K pump also requires both Na^+ (internal) and K^+ (external) for activity; for example, the active extrusion of Na^+ is blocked in zero external K^+. (b) The K^+ concentration required for 50% activation of the enzyme is generally about 1–4 mM, and maximal activation occurs at 6–20 mM. The rate of Na^+ - K^+ pumping is similarly affected by the $[K]_o$ levels. (c) The Na^+ concentration required for 50% activation of the enzyme is generally about 10–20 mM, and maximal activation occurs at 25–130 mM. The active Na^+ extrusion rate is also a function of $[Na^+]_i$. (d) The (Na-K)-ATPase is blocked by cardiac glycosides; half-maximal inhibition usually occurs between 10^{-7} and 10^{-6} M ouabain, and complete inhibition occurs at 10^{-4} to 10^{-3} M. The Na^+-K^+ pump is also blocked by ouabain, with a virtually identical dose-response curve.

(e) Li^+ ion is a poor substitute for Na^+ to cause activation of the enzyme at the inside site. Li^+ also cannot be actively extruded by the cell. (f) Other characteristics of the enzyme and pump are similar, e.g., their pH sensitivities and temperature sensitivities. A 50% depression of the enzyme occurs at a pH of about 6.2 (Sperelakis & Lee, 1971). (g) The (Na-K)-ATPase is found in cell membranes in which Na^+-K^+ pump activity is found. (h) Cells that exhibit a high degree of Na^+-K^+ pumping also possess a high activity of the enzyme. (i) High concentrations of local anesthetics inhibit the enzyme (Henn & Sperelakis, 1968; Sperelakis & Lee, 1971), and local anesthetics also inhibit active Na^+ transport in frog skin (Skou & Zerahn, 1959) and red blood cells (Andersen & Gravenstein, 1965). In general, the hydrolytic activity of the isolated enzyme appears to closely parallel the Na^+-K^+ transport activity of the cell.

9. Inhibition by cardiac glycosides

Inhibition of the (Na-K)-ATPase by cardiac glycosides varies with the species. Rat and mouse hearts are the least sensitive to cardiac glycosides, and guinea pig and rabbit exhibit intermediate sensitivity; in contrast, human, bovine, canine, and feline hearts are most sensitive. The sensitivity of the Na^+-K^+ pump to cardiac glycosides also varies with the species, and provides further evidence that the enzyme and the cation pump are one and the same.

In the most sensitive species, the affinity constant of the cardiac glycoside receptor on the myocardial (Na-K)-ATPase for ouabain is about 2.9×10^8 M^{-1} at 2 mM K^+ and about 1.6×10^8 M^{-1} at 6 mM K^+; this example illustrates the known antagonism of ouabain action by K^+. A 50% inhibition of the enzyme by ouabain usually occurs at 10^{-7} to 10^{-6} M. Current thinking is that the binding site for cardiac glycosides on the (Na-K)-ATPase may be the same receptor that is responsible for the therapeutic and toxic effects of digitalis on the heart.

There is controversy as to whether extremely low concentrations of cardiac glycoside can actually stimulate the (Na-K)-ATPase and whether this stimulation might have any relevance to the mechanism of the therapeutic action of digitalis (see review by Schwartz et al., 1975). Chronic exposure of cultured cells (HeLa) to low doses of ouabain, however, leads to an increase in the number of specific binding sites for [^{3}H]ouabain, and this effect is prevented by inhibitors of protein synthesis (Boardman et al., 1975). It was reported that low concentrations of cardiac glycosides actually stimulate the inward Ca^{2+} current in vascular smooth muscle cells, and that this might explain the coronary artery vasospasms sometimes produced (Belardinelli et al., 1979).

10. Other inhibitors

The (Na-K)-ATPase is inhibited by sulfhydryl reagents, such as *N*-ethylmaleimide, indicating that SH groups are crucial for activity. For example, mercurial diuretics and ethacrynic acid inhibit the isolated (Na-K)-ATPase, and this effect is generally thought to be responsible for the diuretic action of these agents [see 1975 review by Schwartz et al.].

Local anesthetics also inhibit the enzyme (Sperelakis & Henn, 1970). Diphenylhydantoin, however, does not stimulate or inhibit (Sperelakis & Henn, 1970). Oligomycin, an inhibitor of oxidative phosphorylation in mitochondria, inhibits the (Na-K)-ATPase in a wide variety of tissues (Jobsis & Vreman, 1973) and inhibits Na/Na exchange (Garrahan & Glynn, 1967a). Chlorpromazine, a tranquilizer and antidepressant, has only a small inhibitory effect on the (Na-K)-ATPase, although the free radical of chlorpromazine exerts a potent inhibitory action [see Schwartz et al. (1975) for references]. Phlorizin, an inhibitor of Na^+-dependent glucose transport, also inhibits the (Na-K)-ATPase. Ethanol inhibits the (Na-K)-ATPase in vitro, but chronic administration of ethanol to animals leads (adaptively) to an increased (Na-K)-ATPase activity. Aldosterone appears to induce the synthesis of (Na-K)-ATPase molecules in renal tubules and in frog skin; this effect is prevented by pretreatment with inhibitors of protein synthesis.

11. Phosphorylated intermediate

A phosphorylated intermediate of the (Na-K)-ATPase has been found (Post et al., 1965; Skou, 1971). Its phosphorylation by ATP is Na^+ dependent, and its dephosphorylation is K^+ dependent. Cardiac glycosides inhibit the enzyme by attaching to a binding site on the outer surface of the membrane. Although ouabain probably does not bind to the K^+ transport site, there is an exterior ion site to which, if K^+ is bound, ouabain does not bind. Na^+ binding to the exterior ion site does not prevent ouabain binding (Lindenmayer & Schwartz, 1973). Formation of the phosphoenzyme also requires Mg^{2+} (1 mM). The following sequence of reactions has been suggested (Skou, 1975)

$$E + Mg^{2+} \leftrightarrow \text{Mg-E} \qquad \text{(Rx 1)}$$

$$\text{Mg-E} + \text{ATP} \xleftrightarrow{Na^+} \text{Mg-E}_1 \sim \text{P} + \text{ADP} \qquad \text{(Rx 2)}$$

$$\text{Mg-E}_1 \sim \text{P} \leftrightarrow \text{Mg-E}_2\text{-P} \qquad \text{(Rx 3)}$$

$$\text{Mg-E}_2\text{-P} + \text{HOH} \xrightarrow{K^+} \text{Mg-E} + \text{P}_i \qquad \text{(Rx 4)}$$

where E is the enzyme, E_1~P is the high-energy phosphorylated form, and E_2-P is the low-energy phosphorylated form. The phosphate group is bound to a β-carboxyl group of an aspartic acid residue by an acyl bond (Degani & Boyer, 1973; Nishigaki et al., 1974). Because hydroxylamine induces the hydrolysis of the phosphorylated intermediate (i.e., has a K^+-like action), it was concluded (Hokin et al., 1965) that the phosphorylated intermediate was an acyl phosphate. There is some evidence (see Skou, 1975) that the enzyme may have two Mg^{2+} binding sites: (a) a high-affinity site necessary for the formation of the phosphorylated intermediate, and (b) a lower-affinity site necessary for conversion of the high-energy form into the low-energy form (Reaction 3). In the presence of Na^+ but without K^+, a reaction of the enzyme with ATP and Mg^{2+} leads to phosphorylation, with a low rate of dephosphorylation and hydrolysis of ATP; the addition of K^+ leads to a considerable increase in ATP hydrolysis. ITP, GTP, UTP, and CTP all phosphorylate the enzyme at a much slower rate than does ATP. These findings agree with the specific requirement for ATP by the Na^+-K^+ pump.

12. Activator sites

The Na^+ and K^+ activator sites on the enzyme may also be the carrier sites for active transport of these two cations (Skou, 1975). The Na^+ and K^+ affinity characteristics for activation are identical to those for active transport (Post et al., 1960), and the cation binding sites can undergo changes in Na^+-K^+ affinity; these properties thus fulfill one requirement for a carrier site. The sigmoidal shape of the curves for enzyme activation by Na^+ in the presence of K^+, and by K^+ in the presence of Na^+, suggests that more than one cation is necessary for the activation at each site (Skou, 1975). In red blood cells, 3 Na^+ ions are transported out and 2 K^+ ions in per molecule of ATP hydrolyzed (Glynn, 1968). This evidence is most compatible with the view (Skou, 1975) that the inside Na^+-binding site (i-site with n ions per site or n sites with 1 Na^+/site) and the outside K^+-binding site (o-site with m ions per site or m sites with 1 K^+/site) exist simultaneously and independently of each other. The Na^+/K^+ affinity ratios may normally be about 4 : 1 for the inside site and about 1 : 100 for the outside site. Several other cations can replace K^+ on the o-site, the order of effectiveness being: $K^+ > Rb^+ > NH_4^+ > Cs^+ > Li^+$ (Skou, 1965).

Assuming that each site is occupied by either Na^+ or K^+, the enzyme molecules must consist of a mixture of four forms in equilibrium (Skou, 1975):

$$\frac{\text{Outside site:}}{\text{Inside site:}} \quad \frac{Na_o}{Na_i} \leftrightarrow \frac{K_o}{Na_i} \leftrightarrow \frac{K_o}{K_i} \leftrightarrow \frac{Na_o}{K_i} \qquad \text{Rx 5}$$

Only the K_o/Na_i form (second from left) can catalyze the hydrolysis of ATP. The Na_o-Na_i exchange form (left) requires ATP, although the ATP is not hydrolyzed. At given Na-K concentration ratios inside and outside the cell, a decrease in the ratio of [ATP] : [ADP][P_i] shifts the equilibrium from a K_o/Na_i form (Na_i-K_o exchange) toward the Na_o/Na_i form (Na_i-Na_o exchange). This shift gives a decrease in the Na^+ efflux that is coupled to a K^+ influx, and an increase in the Na^+ efflux that is coupled to a Na^+ influx. The K_o/K_i exchange in red blood cells requires ATP, although the ATP is not hydrolyzed (Glynn et al., 1971).

The Na_o/K_i form (right) represents a reversal of the pump; this reversal can occur when the outside solution contains high Na^+ and 0 K^+, and when the inside solution contains high K^+ and 0 Na^+. With pump reversal, *ATP is synthesized* from ADP plus P_i (Garrahan & Glynn, 1967b), about 2–3 K^+ ions being transported outward per molecule of ATP synthesized (Glynn et al., 1971).

All four of these forms of the (Na-K)-ATPase are inhibited by cardiac glycosides, as expected if all four forms represent one basic transport system. Thus, the system has a high degree of flexibility.

There is an efflux of Na^+ that occurs without either K^+ or Na^+ in the external medium (Baker, 1964; Garrahan & Glynn, 1967a), i.e., a **nonexchanging Na^+ pump**.

13. Mechanism of translocation

The mechanism for translocation of the cations is still unknown. ATP changes the (Na-K)-ATPase in such a way that the enzyme leads to an exchange of the cations between the inside site and outside site. Because the Na-K affinity of each site is influenced by ATP, the affinity of each site undergoes a cyclic change due to the reaction with ATP and its subsequent hydrolysis. Skou (1975) suggested that the vectorial transport of Na^+ from the i-site to the o-site, and of K^+ in the opposite direction, is due to an exchange that results from a shift of the i-site from a Na^+ to a K^+ affinity, and of the o-site from a K^+ to a Na^+ affinity. The input of energy from ATP transforms the affinities from outside K^+/inside Na^+ to outside Na^+/inside K^+. Without ATP, the affinities of both the o-site and i-site are toward K^+, and ATP shifts the affinities of both sites toward Na^+. The affinity of the o-site, however, is changed only to a small extent, so that the affinity for K^+ is still almost 100 times higher than that for Na^+. In contrast, at

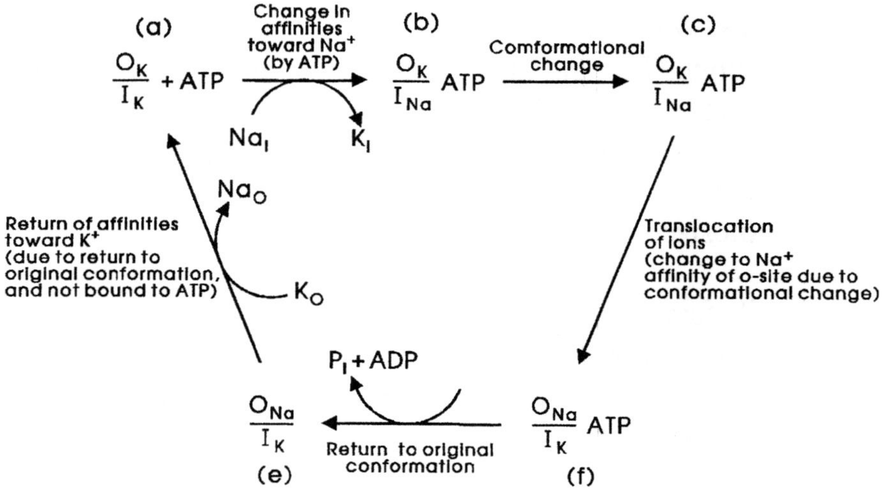

FIGURE II-3. Scheme for the active transport of Na^+ and K^+, as envisioned by Skou. O_K and O_{Na} , -site of K^+ and Na^+, respectively; I_K and I_{Na} , K^+ and Na^+ current, respectively; Na_i and K_i , inward Na^+ and K^+, respectively; Na_o and K_o outward Na^+ and K^+, respectively. [Adapted from Skou (1975).]

the i-site, the ATP shifts the affinity from a K > Na state to a Na > K state; because of this shift, incoming K^+ on the i-site can be exchanged for outgoing Na^+. Upon a conformational change, the Na > K state of the i-site shifts back to a K > Na state. If the energy from hydrolysis of ATP transforms the o-site into a conformation in which the affinity for Na > K, the cations could thus be exchanged between the outer and inner sites. Thus, the appropriate conformational changes of the enzyme may be brought about by phosphorylation and dephosphorylation. The various steps envisioned by Skou (1975) are summarized in Figure II-3 in modified form.

To recapitulate, because Na^+ is required at the i-site for enzyme activation, and because K^+ competes with Na^+ for this site, the fraction of the enzyme molecules that has the i-site in the Na form depends on the Na-K concentration ratio in the internal solution and on the Na-K affinity ratio of the site, which depends on the ATP concentration. Without ATP, the affinity for Na^+ is several times lower than that for K^+, whereas in the presence of a high concentration of ATP, the Na^+ affinity is several times higher than that for K^+. Although ATP also has an effect on the K/Na affinity ratio at the o-site (which must be occupied by K^+ for enzyme activity), even in the presence of ATP, the affinity for K^+ is still about 100 times higher than that for Na^+. The effect on the affinity ratios is relatively specific for ATP: ITP has practically no effect, and CTP and GTP effects require much higher concentrations than for ATP. ADP is almost as potent as ATP, but AMP has no effect (Skou, 1974). The effect is due to ATP as such, and not to a phosphorylation by ATP or to a hydrolysis of ATP.

Additional details about the (Na,K)-ATPase, including molecular structure, are given in Chapter IX, which deals with electrogenic Na^+ pump potentials and measurement of pump current.

Chapter III

Ca^{2+}/Na^{+} EXCHANGE REACTION

A. Energetic Considerations

It was demonstrated in 1969a by Baker et al. that an exchange reaction occurs across the cell membrane of squid giant axon, a reaction that exchanges internal Ca^{2+} for external Na^{+}. Independently, Reuter and Seitz (1968) demonstrated that the exchange reaction was present in cardiac muscle and probably also in some smooth muscles. The exchange reaction usually exchanges 1 Ca^{2+} for 3 Na^{+} ions, although stoichiometries of 1 : 2 or 1 : 4 have been suggested, for example, in squid axon (Blaustein, 1976a,b; Reuter & Seitz, 1968). If the stoichiometry were 1 : 2 , the exchange would be electroneutral. It is assumed that a "carrier" molecule participates in the exchange, either rotating between the two surfaces of the membrane, or undergoing a conformational change that makes the outer cation-binding site become inside and the inner site become outside.

It is proposed that before the carrier can move, binding sites for Na^{+} and Ca^{2+} must be occupied; i.e., the carrier must be loaded before movement can occur. In the rotating carrier version, the carrier is free to rotate when loaded with either 1 Ca or 3 Na at both ends. The exchange carrier does not have much affinity for Li^{+}, K^{+}, Mg^{2+}, or La^{3+} ions, but the affinity for Sr^{2+} is about the same as that for Ca^{2+}. Obviously, other models that are equally feasible can be made.

The molecular structure proposed for the exchanger molecule is illustrated in Figure III-1. As depicted, there are eleven membrane-spanning segments, each containing an alpha-helix structure, and hooked together by connecting links

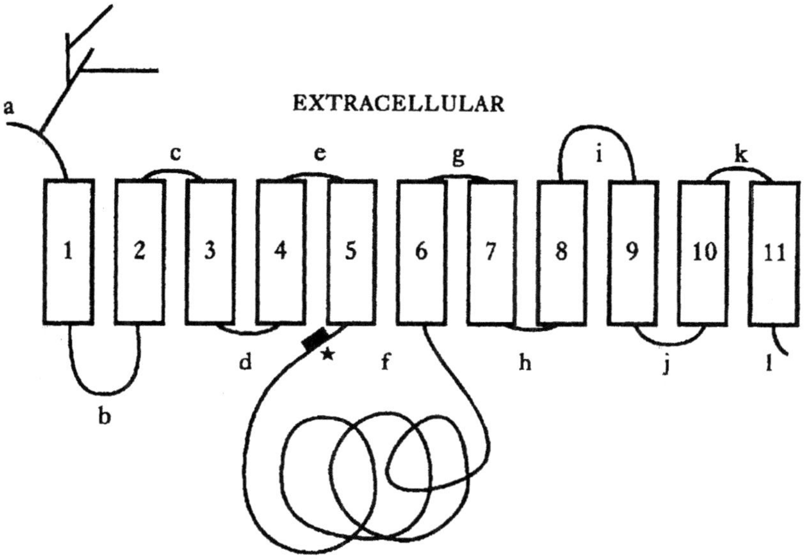

FIGURE III-1. Model for the Na/Ca exchanger based on hydropathy analysis. There are 11 putative membrane-spanning segments (or domains), with consecutive intracellular/extracellular connecting links (labelled *b* to *k*). Each membrane-spanning domain contains an alpha-helix structure. There is a glycosylation site on the extracellular terminus (labelled *a*), and a calmodulin binding site near the beginning of the large intracellular loop (labelled *f*). [Reproduced from K.D. Philipson (1992) with permission.]

that are alternately intracellular and extracellular. The intracellular link between segments (domains) 5 and 6 is very long, and there is a calmodulin-binding site near its beginning. There is a glycosylation site on the extracellular terminus leading to segment 1.

The Ca/Na exchange carrier can, in principle, exchange any of the following combinations: (a) internal Ca^{2+} for external Na^+ (**forward mode**), (b) external Ca^{2+} for internal Na^+ (**reverse mode**) (Fig. III-2), (c) internal Ca^{2+} for external Ca^{2+}, and (d) internal Na^+ for external Na^+. The Ca/Na or Na/Na exchange could account for the well-known ouabain-insensitive Na^+ efflux (fraction of total Na^+ efflux), which is markedly reduced when the external Ca^{2+} concentration is zero and is depressed when the external Na^+ concentration is low (Baker, 1964). The Ca/Na exchange is probably the important exchange, the forward mode (Ca_i/Na_o) being the most important physiologically in cells at rest, and the reverse mode (Ca_o/Na_i) being important in excited cells and under some pathological conditions or drug actions when $[Na]_i$ is abnormally high. What is exchanged for what depends on the relative concentrations of Ca^{2+} and Na^+ on each side of the membrane, and on the relative affinities of the binding sites on the carrier for Ca^{2+} and Na^+. The relative affinities for Ca^{2+} compared to Na^+ could vary, depending on which surface of the membrane that the carrier is loading (or unloading). The idea is that Ca^{2+} and Na^+ compete with each other for binding to the carrier, and that the relative affinities and conditions inside favor Ca^{2+} binding and those outside favor Na^+ binding. Thus, under normal conditions in resting cells, the internal Ca^{2+} is exchanged for external Na^+ (Fig. III-2).

This exchange effectively acts somewhat like a Ca^{2+} pump directed outward, a pump whose energy is derived from the Na^+ electrochemical gradient (directed inward), which is

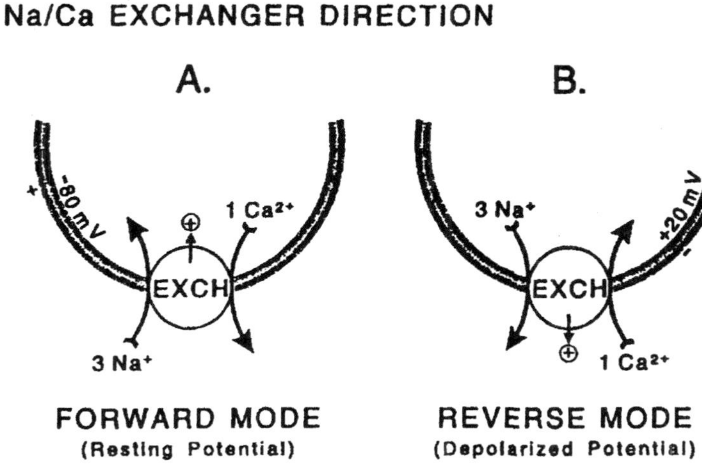

FIGURE III-2. The electrogenic Na/Ca exchange depicted in the forward mode (A) and reverse mode (B) of operation for a coupling ratio of 3 Na^{+} : 1 Ca^{++}. In the forward mode, 3 external Na^{+} ions are exchanged for one internal Ca^{2+} ion. Thus, the net exchanger current ($I_{Na/Ca}$) is *inwards*, which is *depolarizing*. In the reverse mode, 3 internal Na^{+} ions are exchanged for one external Ca^{2+} ion; thus, $I_{Na/Ca}$ is *outwards* and *hyperpolarizing*.

ultimately driven by the Na^+-K^+ pump (Na-K)-ATPase. Under some conditions during which $[Na]_i$ is elevated, e.g., under action of cardiac glycosides that inhibit the (Na-K)-ATPase, the higher internal Na^+ can more effectively compete with the internal Ca^{2+} for binding to the exchange carrier. Thus, less Ca^{2+} is transported out, and consequently $[Ca]_i$ increases; this increase could exert a significant positive inotropic effect. Indeed, such a mechanism has been proposed for the positive inotropic action of cardiac glycosides (Langer, 1968; Glitsch et al., 1970). Although metabolic inhibitors also produce increases in $[Na]_i$ and $[Ca]_i$, but without a transient period of positive inotropy, the concomitant lowering of the cellular level of ATP may effectively counter-balance the Ca^{2+} effect.

As stated in the preceding paragraph, the energy for this apparent Ca^{2+} pump comes from the Na^+ electrochemical gradient maintained by the Na^+-K^+ pump at the expense of ATP. The energy required to pump out Ca^{2+} is about

$$\begin{aligned} \Delta G_{\mathrm{Ca}} &= zF(E_m - E_{\mathrm{Ca}}) \\ &= (+2)\,(10^5)\,[-0.080-(+0.129)] \\ &= (2\times 10^5\ \mathrm{coul/mol})\,(-0.209\ \mathrm{V}) \\ &= 41{,}800\ \ \mathrm{joul/mol} \end{aligned} \qquad \text{eq. 1}$$

E_{Ca} is estimated to be +129 mV, which is calculated from:

$$E_{\text{Ca}} = \frac{-RT}{zF} \ln \frac{[\text{Ca}]_i}{[\text{Ca}]_o} \qquad \text{eq. 2}$$

$$= -2.303 \frac{RT}{zF} \log \frac{[\text{Ca}]_i}{[\text{Ca}]_o}$$

$$= \frac{-60 \text{ mV}}{z} \log \frac{[\text{Ca}]_i}{[\text{Ca}]_o}$$

$$= \frac{+60 \text{ mV}}{+2} \log \frac{[\text{Ca}]_o}{[\text{Ca}]_i}$$

$$= +30 \text{ mV} \log \frac{2 \times 10^{-3} \text{ M}}{1 \times 10^{-7} \text{ M}}$$

$$= +129 \text{ mV}$$

assuming the free internal Ca^{2+} concentration is 1×10^{-7} M. On the other hand, the free internal Ca^{2+} rises to about 10^{-5} M during contraction, and E_{Ca} then becomes only about +69 mV:

$$E_{\text{Ca}} = +30 \text{ mV} \log \left[\frac{2 \times 10^{-3} \text{ M}}{1 \times 10^{-5} \text{ M}}\right] \qquad \text{eq. 2a}$$

$$= +30 \text{mV} \log (2 \times 10^{2})$$

$$= +30 \text{mV} (2.30)$$

$$= +69 \text{mV}$$

In this example, the energy required to pump out Ca^{2+} is 29,800 joul/mol (assuming E_m is -80 mV), but would be only 9,800 joul/mol if E_m is at +20 mV (peak of action potential plateau). The comparable calculations for the energy available from the Na^+ electrochemical gradient are

$$\Delta G_{\text{Na}} = zF\,(E_m - E_{\text{Na}}) \qquad \text{eq. 3}$$

$$= (+1)\,(10^5)\,[-0.080 - (+0.060)]$$

$$= (1 \times 10^5 \text{ coul/mol})\,(-0.140 \text{ V})$$

$$= 14{,}000 \text{ joul/mol}$$

E_{Na} is estimated to be +60 mV:

$$E_{Na} = \frac{+60\ mV}{z} \log \frac{[Na]_o}{[Na]_i} \qquad \text{eq. 4}$$

$$= \frac{+60\ mV}{+1} \log \frac{[Na]_o}{[Na]_i}$$

$$= +60\ mV \log \frac{150 \times 10^{-3}\ M}{15 \times 10^{-3}\ M}$$

$$= +60\ mV$$

assuming the free internal Na^+ concentration in the myoplasm to be 15 mM. Therefore, movement of two mol of Na^+ down its gradient (if 2 Na^+ are exchanged for 1 Ca^{2+}) could theoretically provide up to a maximum of 28,000 joul in a resting membrane. This amount is not enough energy for the 41,800 joul required for 1 mol of Ca^{2+} to be transported out in a resting cell. If the stoichiometry were 3 : 1, as it may be for squid axon, then the Na^+ electrochemical gradient could provide sufficient energy (Blaustein, 1976a).

If the free cytoplasmic Na^+ were much lower than 15 mM, e.g., 1 mM, then the energy available from the Na^+ gradient would be considerably greater: E_{Na} would be +131 mV, and ΔG_{Na} would be 21,100 joul/mol. This amount is just about enough to provide the required energy for the Ca^{2+} transport: 2 mol of Na^+ would provide 42,000 joul, compared to the required energy of 41,800 joul for 1 mol of Ca^{2+}. However, Lee & Fozzard (1975) reported an internal Na^+ activity of 5.7 mM for myocardial cells.

The discussion above applies to the resting membrane. During the action potential (AP) (e.g., cardiac plateau), the energy required for Ca^{2+} transport is less, but so is the energy available from the Na^+ gradient. There is one difference, however: during excitation the relative increase in the free $[Ca]_i$ (from 10^{-7} M to about 10^{-5} M) is much greater (because of the lower absolute levels) than that for $[Na]_i$ (about 1–2 mM). This change would act to make the energetics more favorable for the Ca/Na exchange reaction in the forward mode of operation during excitation. If $[Ca]_i$ did not increase, then depolarization to the plateau level would switch the exchanger into the reverse mode. Table III-1 summarizes the values for the different conditions.

In the Ca/Na exchange reaction, uphill transport of Ca^{2+} is coupled with downhill transport of Na^+, and the energetics should be as discussed in the preceding paragraphs. That is, a Ca^{2+} electrochemical gradient can be established across a membrane as a consequence of an existing Na^+ gradient. In such a system at equilibrium, if only two univalent Na^+ ions are transported for each divalent Ca^{2+} ion, the distribution ratios that would occur are (see Reuter, 1974)

$$\frac{[Ca^{2+}]_i}{[Ca^{2+}]_o} = \frac{[Na^+]_i^2}{[Na^+]_o^2} \qquad \text{eq. 5}$$

The derivation of this relationship is given in the Appendix to this chapter. Then, for a resting cell

$$\frac{1\times10^{-7}\ \text{M}}{2\times10^{-3}\ \text{M}} = \frac{(1\times10^{-3}\ \text{M})^2}{(0.150\times10^{0}\ \text{M})^2} \qquad \text{eq. 5a}$$

$$0.5\times10^{-4} \cong 0.44\times10^{-4}$$

Thus, this is another method of demonstrating that the energetics would be adequate for this uphill-downhill coupled transport system, if cytoplasmic free $[Na]_i$ were to be only about 1 mM. If not, energy must be supplied by some means.

TABLE III-1

Calculations of the energetics of the Ca-Na exchange reaction for several different conditions.

	Ca^{2+}				Na^{+}					
E_m mV	$[Ca]_o$ M	$[Ca]_i$ M	E_{Ca} mV	Energy required ΔG_{Ca} joul/mol	$[Na]_o$ mM	$[Na]_i$ mM	E_{Na} mV	Energy Available ΔG_{Na} joul/mol for 1 Na^+	for 2 Na^+	for 3 Na^+
-80	2×10^{-3}	1×10^{-7}	+129	41,800	150	15	+60	14,000	28,000	42,000
-80	2×10^{-3}	1×10^{-5}	+69	29,800	150	1	+131	21,100	42,200	63,300
+20	2×10^{-3}	1×10^{-5}	+69	9,800	150	1	+131	11,100	22,200	33,300
+20	2×10^{-3}	1×10^{-7}	+129	21,800	150	15	+60	4,000	8,000	12,000

Change in free energy required for Ca^{2+} translocation. Change in free energy available from Na^+ distribution and translocation. Resting potential assumed to be -80 mV, and average potential during the action potential plateau to be +20 mV. Resting $[Ca^{2+}]_i$ assumed to be 1×10^{-7} M, and that during the plateau to be 1×10^{-5} M. The reaction presumably exchanges 1 internal Ca^{2+} ion for 3 external Na^+ ions. The calculations indicate that the energetics are not favorable for 1 Ca^{++} : 2 Na^+ exchange unless $[Na^+]_i$ is close to 1 or 2 mM. If the stoichiometry were 3 Na : 1 Ca, however, then sufficient energy would be available from the Na^+ electrochemical gradient, even if $[Na]_i$ were 15 mM.

Jundt and Reuter (1976) have reported that the Na-activated Ca efflux from guinea pig auricles is partly depressed in metabolically depleted muscle (2 mM cyanide, zero glucose; ATP one-tenth of normal), the depression thus suggesting that a fraction of the Na-dependent Ca efflux was dependent on ATP, as in squid axon (Baker & Glitsch, 1973; DiPolo, 1973). It has been suggested that ATP may facilitate the exchange by increasing the affinities of the carrier for Na and Ca ions (Baker et al., 1969a; Baker & Glitsch, 1973; Jundt et al., 1975; Blaustein, 1976b).

Rearranging Equation 5 gives

$$\frac{[Ca^{2+}]_i}{[Na^+]_i^2} = \frac{[Ca^{2+}]_o}{[Na^+]_o^2} \qquad \text{eq. 5b}$$

which is the familiar $[Ca^{2+}]_o/[Na^+]^2_o$ ratio of Wilbrandt and Koller (1948) and Luttgau and Niedergerke (1958) for control of cardiac muscle contraction. Equation 5b illustrates that the internal Ca^{2+} concentration, and hence the tension, is increased if $[Ca]_o$ is increased or if $[Na]_o$ is decreased (presuming that $[Na]_i$ is held constant).

In addition to explaining the well-known antagonistic effect of Ca^{2+} and Na^+ on the contractility of cardiac muscle (Luttgau & Niedergerke, 1958), the Ca/Na exchange reaction can also explain why: (a) A large fraction of the Ca^{2+} efflux is dependent on external Na^+ and Ca^{2+}. The increase in Ca^{2+} efflux produced by caffeine and metabolic poisons (Ca^{2+} release from the SR into cytoplasm) is also dependent on the external Na^+ and Ca^{2+} concentrations (Nayler, 1963; Jundt et al., 1975). (b) Ca^{2+} efflux is not suppressed by cardiac glycosides, and cyanide and other metabolic inhibitors actually increase Ca efflux (Reuter & Seitz, 1968; Blaustein &

Hodgkin, 1969; DiPolo, 1973; Jundt et al., 1975). (c) The temperature coefficient for Ca^{2+} efflux (Q_{10} of about 1.35) was lower than expected for an active transport system. These and other arguments are presented in a review article by Reuter (1974).

Besides the cell membrane, this Ca/Na exchange reaction could also occur across other membranes, such as across the SR membrane. If so, and if the lumen of the SR contains high Na^+ (as seems to be true for the SR of skeletal muscle), then the energy of the Ca^{2+} gradient can be used to pump Na^+ into the SR or vice versa; i.e., the Na^+ gradient could be used to pump Ca^{2+} into the SR for relaxation. The SR of cardiac and skeletal muscle contains a ouabain-sensitive ATPase activity, presumably (Na-K)-ATPase (Forbes & Sperelakis, 1972). It is possible, in addition, that Na^+ influx into the cytoplasm during the cardiac action potential (AP) could act to release Ca^{2+} from the SR for contraction by the same Ca/Na exchange reaction. Reuter (1974) and Jundt and colleagues (1975), however, contend that the main function of the Na/Ca exchange across the sarcolemma is to serve as a backup mechanism to the internal Ca storage system (the SR) in order to avoid overloading the myocardial cell with Ca.

B. Ratio of Free Energy Changes

The ratio of the free energy changes for Na^+ to Ca^{++} were calculated for coupling ratios of 3 Na^+: 1 Ca^{++} and 4 Na^+: 1 Ca^{++} (3 $\Delta G_{Na}/\Delta G_{Ca}$ and 4 $\Delta G_{Na}/\Delta G_{Ca}$) and plotted as a function of membrane potential, and for different $[Ca]_i$ levels and $[Na]_i$ levels (Figs. III-3 - III-6). This was done to allow a simple qualitative assessment of how the directionality of the exchanger is affected by E_m ; i.e., *forward mode* and *reverse mode* of operation, and how the **equilibrium potential (reversal potential)** of the exchanger, $E_{Na/Ca}$, is shifted by

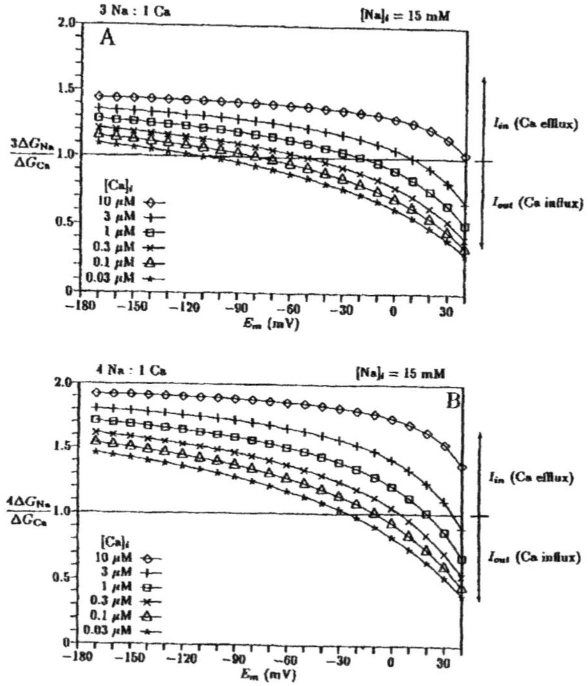

Fig. III-3. Electrogenic Na/Ca exchange for $[Na]_i$ of 15 mM. Calculated ratio of the free energies for Na^+ versus Ca^{++} ($3\Delta G_{Na}/\Delta G_{Ca}$ or $4\Delta G_{Na}/\Delta_{Ca}$) plotted as a function of membrane potential (E_m) for coupling ratios of 3 Na^+: 1 Ca^{++} (A) and 4 Na^+: 1 Ca^{++} (B) for six different intracellular Ca^{++} concentrations ($[Ca]_i$). At a ratio of 1.0, 3 $\Delta G_{Na} = -\Delta G_{Ca}$ and the sum equals zero: 3 $\Delta G_{Na} + \Delta G_{Ca} = 0$. Therefore, the exchange would be at equilibrium at the E_m value at which each $[Ca]_i$ curve crosses the ratio of 1.0 line, i.e., this gives the value of the exchanger equilibrium potential, $E_{Na/Ca}$. At ΔG ratios > 1.0, there is a net *inward current* carried by Na^+ ion, coupled with net *Ca^{++} efflux* from the cell. That is, this represents the **forward mode** of operation of the exchanger. At ΔG ratios < 1.0, there is a net *outward current* carried by Na^+ ion, coupled with net *Ca^{2+} influx* into the cell. This reflects the **reverse mode** of operation of the exchanger. Note that at the higher (more positive) E_m levels, the ΔG ratio is reduced. Also note that at the higher $[Ca]_i$ levels, there is a right-ward shift of $E_{Na/Ca}$. The calculations in this figure and the subsequent five figures were made by Humberto Ortiz-Zuazaga.

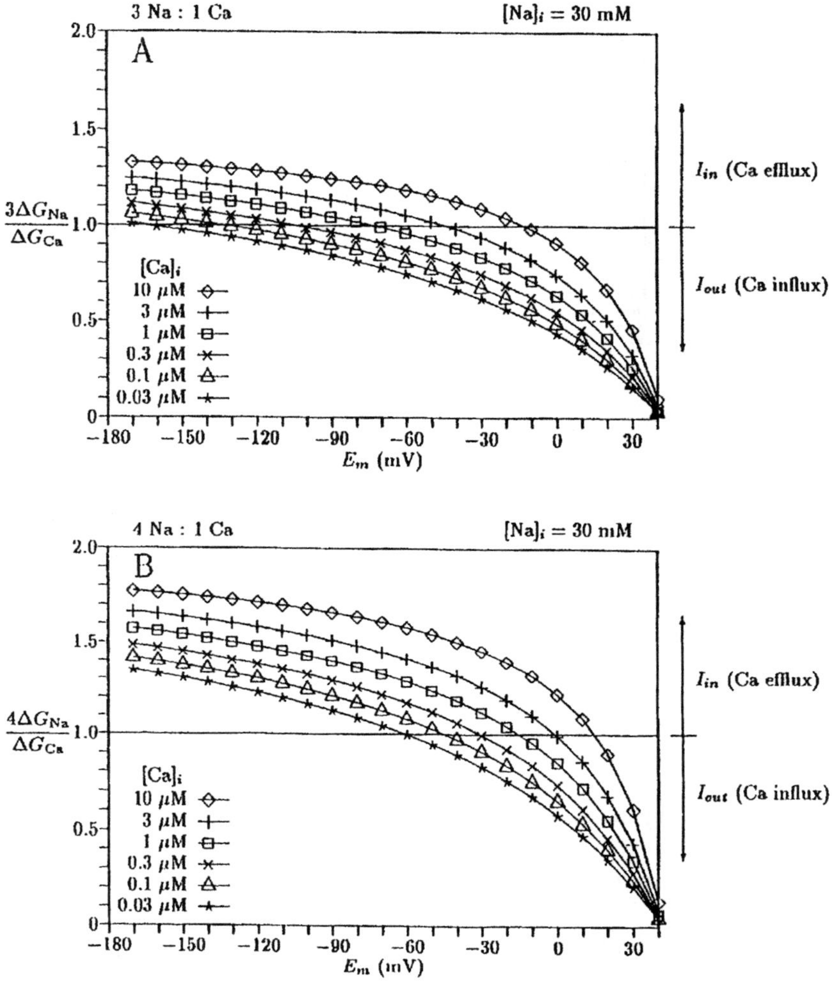

Fig. III-4. Electrogenic Na/Ca exchange for $[Na]_i$ of 30 mM, and a coupling ratio of 3 Na: 1 Ca (panel **A**) and 4 : 1 (panel **B**). Note the upward and rightward shift of the curves at 4 : 1. See legend to Figure III-3 for additional details.

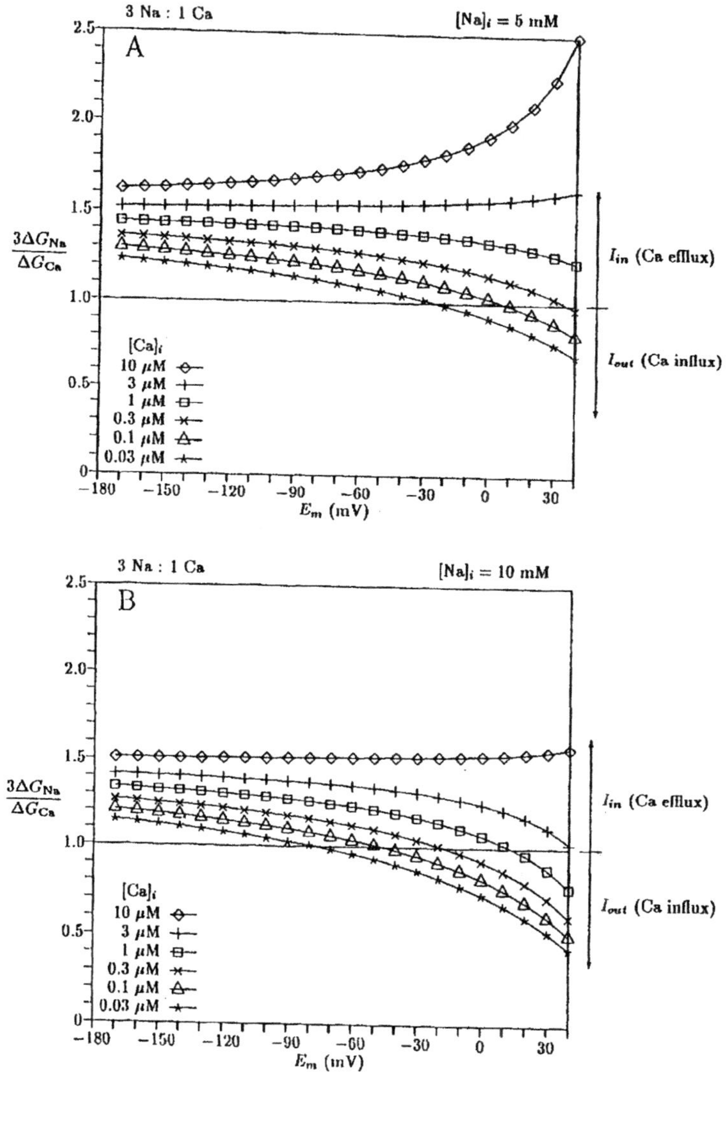

Fig. III-5. Electrogenic Na/Ca exchange for $[Na]_i$ of 5 mM (panel A) and 10 mM (panel B), at a coupling ratio of 3 Na^+ : 1 Ca^{++}. See legend to Figure III-3 for additional details.

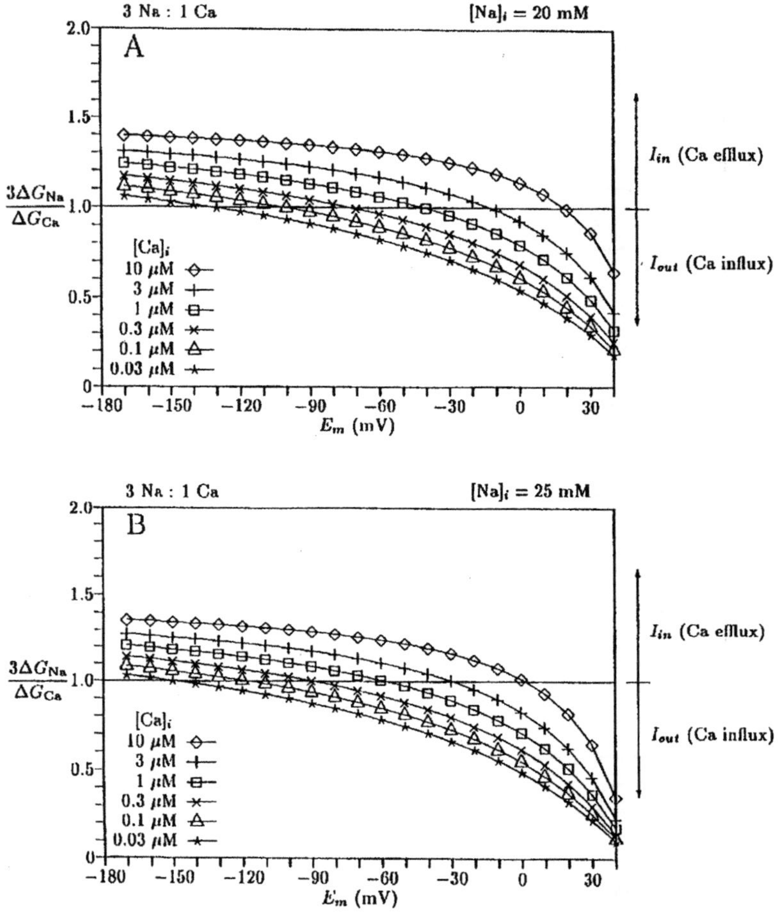

Fig. III-6. Electrogenic Na/Ca exchange for $[Na]_i$ of 20 mM (panel A) and 25 mM (panel B), at a coupling ratio of 3 Na^+ : 1 Ca^{++}. See legend to Figure III-3 for additional details.

TABLE III-2

Calculated values for 3 $\Delta G_{Na}/\Delta G_{Ca}$ assuming an exchange ratio of 3:1.

[Na]$_i$ (mM)	[Ca]$_i$ (M) 10^{-7}		10^{-6}		10^{-5}	
	-80	+20	-80	+20	-80	+20 mV
5	1.21	0.94	1.41	1.30	1.70	2.10
15	1.00	0.55	1.17	0.76	1.41	1.22
30	0.86	0.30	1.02	0.42	1.23	0.67

Values given for the assumed resting potential (-80 mV) and AP plateau potential (+20 mV) for several [Ca]$_i$ levels and [Na]$_i$ levels.

TABLE III-3

Calculated values for $E_{Na/Ca}$ assuming an exchange ratio of 3:1.

[Na]$_i$ (mM)	[Ca]$_i$ (μM) 0.03	0.1	0.3	1	3	10
5	-24	+ 8	+36	+69*	+98*	+30*
10	-78	-46	-18	+13	+43*	+75*
15	-109	-78	-49	-18	+10	+43*
20	-132	-101	-72	-41	-12	+19
25	-149	-118	-89	-58	-29	+ 2
30	-164	-132	-104	-72	-44	-12

Values given for several [Ca]$_i$ levels and [Na]$_i$ values.

*These values were calculated from the following equation:

$$E_{Na/Ca} = 3E_{Na} - 2E_{Ca}$$
$$= 3\frac{61}{1}\log\frac{[Na]_o}{[Na]_i} - 2\frac{61}{2}\log\frac{[Ca]_o}{[Ca]_i}$$

$[Ca]_i$ and $[Na]_i$ (assuming constant extracellular concentration of 150 mM Na^+ and 2.0 mM Ca^{++}).# When the ratio is 1.0, 3 $\Delta G_{Na} = -\Delta G_{Ca}$ and the sum equals zero: 3 $\Delta G_{Na} + \Delta G_{Ca} = 0$. Therefore, the exchanger would be at equilibrium. At ΔG ratios > 1.0, there is a net *inward current* carried by Na^+ ion, coupled with net *Ca^{++} efflux* from the cell, representing **forward mode** of operation of the exchanger. At ΔG ratios < 1.0, there is a net *outward current* carried by Na^+ ion, coupled to a net *Ca^{++} influx* into the cell, representing **reverse mode** of operation of the exchanger.

For constant $[Ca]_i$ and $[Na]_i$ levels, the ΔG ratio is reduced at the higher (more positive) E_m levels (Figs. III-3 - III-6). Elevation of $[Ca]_i$ produces a rightward shift of $E_{Na/Ca}$. For example, when a myocardial cell changes from the resting potential (-80 mV) to the AP plateau potential (+20 mV), simultaneous with $[Ca]_i$ being elevated from about 1 x 10^{-7} M to 3 x 10^{-6} M, the exchanger goes into *reverse mode* with Ca^{++} influx (Fig. III-3A). However, if the coupling ratio were 4 : 1, reversal of the exchanger would not occur if $[Ca]_i$ rose to higher than 1.0 μM (Fig. III-3B). The same would be true if $[Na]_i$ were 5 or 10 mM (Fig. III-5). At 5 mM $[Na]_i$, there would be a continual Ca^{++} efflux via the exchanger (*forward mode*) at the physiological potentials and at 0.3 μM $[Ca]_i$ or higher. Conversely, if the $[Na]_i$ level were 25 or 30 mM, the exchanger would always operate in *reverse mode* at the physiological E_m levels (Figs. III-4A, III-6B), unless the coupling ratio were 4 : 1 (Fig. III-4B). The actual values of 3 $\Delta G_{Na}/\Delta G_{Ca}$, for an exchange ratio of 3 : 1, are listed in Table III-2 for three $[Na]_i$ levels and $[Ca]_i$ levels, for two physiological

---

In a letter to the author in 1979, subsequent to publication of the original Handbook article, Dr. Tobias Op't Hof suggested that such plots be made.

potentials, -80 mV (resting potential) and +20 mV (AP plateau potential).

The reversal potential for the Na/Ca exchanger, for a coupling ratio of 3 Na^+ : 1 Ca^{++}, is given by the following equation:

$$E_{Na/Ca} = 3\,E_{Na} - 2\,E_{Ca} \qquad \text{eq. 6}$$

This equation indicates that the reversal potential of the exchanger, $E_{Na/Ca}$, is solely a function of the ion distributions and coupling ratio. For constant $[Na]_o$ and $[Ca]_o$ values of 150 mM and 2.0 mM, respectively, and for a coupling ratio of 3 Na^+ : 1 Ca^{++}, how $E_{Na/Ca}$ would vary with changes in the levels of $[Na]_i$ and $[Ca]_i$ is graphed in Figure III-7. The family of curves in panel **A** is for $[Ca]_i$ levels between 0.03 μM and 10 μM, and illustrates how $E_{Na/Ca}$ shifts to more negative potentials as $[Na]_i$ level is elevated. The absolute value of $E_{Na/Ca}$ is less negative (more positive) when the $[Ca]_i$ level is elevated, e.g., during excitation-contraction coupling.

The family of curves in panel B of Figure III-7 is for $[Na]_i$ levels between 5 mM and 30 mM, and illustrates how $E_{Na/Ca}$ shifts to more positive (less negative) potentials as $[Ca]_i$ is elevated during the action potential. The absolute value of $E_{Na/Ca}$ is less negative (more positive) when the $[Na]_i$ level is lower. Note that all the curves are straight lines on the semilogarithmic plots with parallel slopes, as expected from the Nernst equation for calculation of E_{Na} and E_{Ca}. The actual values of $E_{Na/Ca}$, for an exchange ratio of 3 : 1, are listed in Table III-3 for the different $[Na]_i$ and $[Ca]_i$ levels.

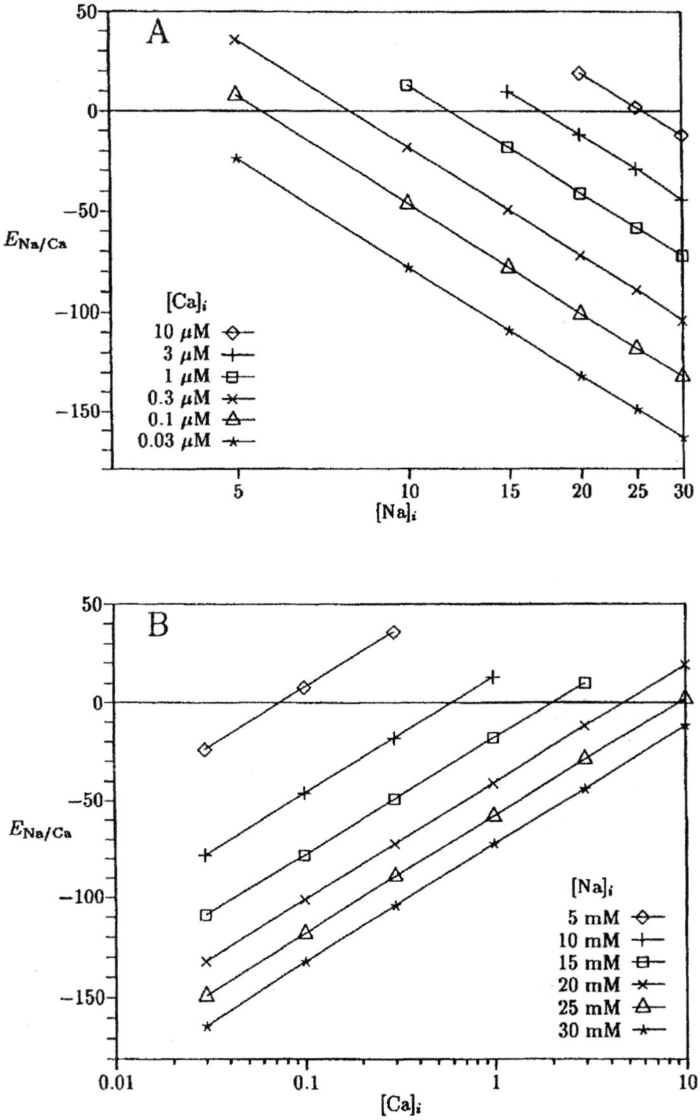

Fig. III-7. The reversal potential ($E_{Na/Ca}$) for the Na/Ca exchanger, assuming a coupling ratio of 3 Na^+ : 1 Ca^{++}, is plotted on the ordinate as a function of $[Na]_i$ (panel A) or $[Ca]_i$ (panel B) on the abscissa. In A, the family of curves is for six different $[Ca]_i$ levels, and in B, for six $[Na]_i$ levels.

C. Electrogenic Na/Ca Exchange Current Measurement

1. General Comments

An illustration of the experimental measurement of the electrogenic Na/Ca exchange current ($I_{Na/Ca}$), and its relationship to the transient inward current (I_{TI}) of myocardial cells, is given in this section for cultured single ventricular myocytes. Whole-cell voltage clamp can be used to record $I_{Na/Ca}$ and I_{TI} currents from the isolated myocytes. Patch pipettes having resistances of 3–5 MΩ can be used, and seal resistances of 2–4 GΩ obtained. The series resistance and the leakage current should be compensated. The input capacitance (C_{in}) of the cells can be determined by applying ramp pulses (-5 mV amplitude; 0.2 V/sec slope), and the capacitative surface area (A_s) of the cell calculated by dividing the measured C_{in} with the assumed specific membrane capacitance (C_m of 1.0 μ F/cm^2). The density of the membrane currents (in μA/cm^2) is obtained by dividing the current amplitude by A_s.

2. Measurement of $I_{Na/Ca}$

The inwardly-directed $I_{Na/Ca}$ (forward mode of the exchanger; 3 Na^+ in : 1 Ca^{2+} out) can be isolated for whole-cell voltage-clamp measurement of $I_{Na/Ca}$ (Kimura et al., 1987). The myocytes can be superfused with an external solution containing (mM) for example: NaCl 143 or 0 (substituted with LiCl), $CaCl_2$ 0.9, $MgCl_2$ 1.0, glucose 5.5, Hepes 5.0 (pH 7.4), TTX 0.01, ouabain 0.01, $BaCl_2$ 2.0, CsCl 2.0, and $CoCl_2$ 2.0. The pipette electrodes may be filled with a solution containing (mM): Cs-glutamate 50, Mg-ATP 10, Na_2-phosphocreatine 5.0, TEA-Cl 20, and HEPES 5.0 (pH 7.4) and pCa 6.36 (Fabiato & Fabiato, 1978). A Na-K pump inhibitor, such as ouabain (or palmityl-carnitine 5.0 μM, a toxic metabolite of ischemia) may be applied in the superfusing solution. Under these conditions, the voltage- and time-dependent Na^+, Ca^{2+}, and K^+ conductances are blocked and the Na-K pump is inhibited.

The Na/Ca exchanger is activated by changing the Na^+-free superfusate to the Na^+-containing solution, and the current/voltage relationship may be obtained by applying ramp voltage pulses from a holding potential (HP) of about -40 mV: an initial sudden depolarization to +60 mV, a gradual hyperpolarization (over 500 msec) to -120 mV, and a sudden return to the HP.

The I/V relationship for $I_{Na/Ca}$ shows a voltage dependence. The measured E_{rev} of $I_{Na/Ca}$ should agree with the equilibrium potential ($E_{Na/Ca}$) predicted for an electrogenic 3 Na^+ / 1 Ca^{2+} exchange. The density of peak $I_{Na/Ca}$ is about 2-4 $\mu A/cm^2$ in control conditions. $I_{Na/Ca}$ is completely blocked by dimethyl-amiloride (20 μM). Increasing $[Ca]_i$ shifts $E_{Na/Ca}$ to more positive potentials and increases $I_{Na/Ca}$

3. Measurement of I_{TI}

The properties of the transient inward current (I_{TI}) also can be studied under whole-cell voltage clamp. A Na-K pump inhibitor can be used to induce arrhythmogenic transient inward current (I_{TI}) in isolated ventricular myocytes. The cardiomyocytes may be superfused with an external solution containing (mM): NaCl 143, KCl 5.4, $CaCl_2$ 2.5, $MgCl_2$ 1.0, glucose 5.5, HEPES 5 (pH 7.4), and TTX 0.01. The pipette solution can contain (mM): K-aspartate 147, $MgCl_2$ 1.0, Mg-ATP 10, Na_2-phosphocreatine 5.0, EGTA 0.1, HEPES 5.0 (pH 7.4). The I_{TI} currents may be elicited by 1 sec depolarizing voltage pulses, in 10 mV steps, to +60 mV from a HP of about -60 mV.

I_{TI} can be completely blocked by 20 μM dimethyl-amiloride or 1 μM ryanodine. The amplitude of I_{TI} is proportional to the magnitude and duration of the conditioning depolarizing voltage pulse. Density of peak I_{TI} is about 7-9 μ A/cm^2 (at -60 mV), and the current gradually decreases at the more-negative E_m values, going to nearly zero at about -140 mV. The I_{TI} current results from the cyclic release of Ca^{2+} from the SR, and appears to be mediated at least in part by the electrogenic $I_{Na/Ca}$ (see below).

4. I_{TI} and $I_{Na/Ca}$

It has been proposed that I_{TI} can be mediated by two mechanisms: (a) a nonselective (Na^+, K^+) cation channel or (b) the electrogenic Na/Ca exchanger. Evidence favoring the nonselective cation channel has come from experiments on Purkinje fibers, whereas observations made on myocardial cells suggest that the Na/Ca exchanger current may be the primary mechanism for production of I_{TI}. It is possible that both mechanisms exist in both Purkinje fibers and myocardial cells, but the mechanism predominating may vary with cell type (for review, see January & Fozzard, 1988; Ferrier, 1991).

The mechanism generally accepted for induction of I_{TI} requires an increase in $[Ca]_i$, which causes overloading of the SR with Ca^{2+}, and induces a spontaneous oscillatory release of Ca^{2+} from SR following depolarizing pulses or an AP (Berlin et al., 1989; Giles & Shimoni, 1989; Katz et al., 1980). The cyclic release of Ca^{2+} from the SR activates either or both (a) Ca^{2+}-activated non-selective cation channels (Colquhoun et al., 1981; Kass et al., 1978; Lederer & Tsien, 1976) and/or (b) the electrogenic Na/Ca exchanger (Arlock & Katzung, 1985; Fedida et al., 1987; Kimura et al., 1987).

There are at least two different mechanisms to cause intracellular Ca^{2+} overload: (a) stimulation of the Ca^{2+} influx into the cell during a long-lasting depolarization via the voltage-dependent Ca^{2+} channel during a long-lasting depolarization, and (b) inhibition of the sarcolemmal (Na^+-K^+)-ATPase activity. Sodium pump inhibition leads to intracellular Na^+ accumulation, and therefore an increase in $[Ca]_i$ resulting from reduced Ca^{2+} efflux through the Na/Ca exchanger (Giles & Shimoni, 1989; January & Fozzard, 1988; Kass et al., 1978; Kojima & Sperelakis, 1984; Meszaros & Pappano, 1990).

The findings that ryanodine, an SR Ca^{2+}-release channel blocker (Marban & Wier, 1985), and dm-Amil, a Na/Ca exchange blocker (Kleyman & Cragoe, 1988), are able to abolish induced I_{TI} indicate a crucial role for the oscillatory release of Ca^{2+} from the SR. Agents that induce I_{TI} , like ouabain, have no significant direct effect on the Na/Ca exchanger, but act indirectly by inhibition of the Na-K pump current. Philipson (1984) reported that charged amphiphiles (like PalCar) do not exert an effect on Na/Ca exchange.

The Na/Ca exchanger can transport Ca^{2+} in both outward (forward mode) and inward (reverse mode) directions, depending on the direction of the electrochemical gradient for Na^+, which is the energy source for this process (Fig. III-2). The Na/Ca exchanger transports 1 Ca^{2+} ion per 3 Na^+ ions, thus generating a net membrane current carried by Na^+ (Mullins, 1979; Eisner & Lederer, 1985). The magnitude of this electrogenic Na/Ca exchange current ($I_{Na/Ca}$) depends on the membrane potential (Kimura et al., 1987; Lipp & Pott, 1988). The measured reversal potential for $I_{Na/Ca}$ agrees with the equilibrium potential ($E_{Na/Ca}$) predicted for 3 Na^+/1 Ca^{2+} exchange (Ehara et al., 1989).

The Na^+ electrochemical gradient energizes Ca^{2+} extrusion from the cell through the Na/Ca exchange, because $E_{Na/Ca}$ is less negative (about -45 mV) than V_m (about -80 mV). During an AP, or a long depolarizing pulse, when V_m becomes less negative, the exchanger reverses and now extrudes Na^+ and brings Ca^{2+} into the cell (Ehara et al., 1989; Kimura et al., 1987).

During Na-K pump inhibition, $[Na]_i$ increases, causing $E_{Na/Ca}$ to shift to more negative (hyperpolarized) potentials. Under these circumstances, the exchanger works in reverse mode, and extrudes Na^+ at the expense of Ca^{2+} gain (Fig. III-8). This situation is further accentuated during depolarization, and the cell becomes overloaded with Ca^{2+}, thus leading to a sufficient rise in $[Ca]_i$ to trigger Ca^{2+} release from the SR. The increasing $[Ca]_i$ shifts $E_{Na/Ca}$ to more positive potentials. If now V_m suddenly jumps back to a negative potential, at the termination of the depolarizing pulse or AP, $E_{Na/Ca}$ remains positive for a while, thus promoting the activation of a large inward $I_{Na/Ca}$ which mediates I_{TI}.

Ouabain induces Ca^{2+}-overload and resultant I_{TI} by inhibiting the (Na^+, K^+)-ATPase, which indirectly activates $I_{Na/Ca}$ via the mechanism described above. PalCar also has been demonstrated to inhibit the (Na^+, K^+)-ATPase (Abe et al., 1984; Adams et al., 1979; Wood et al., 1977) and to block the electrogenic Na-K pump current in ventricular myocytes (Pappano et al., 1990; Tanaka et al., 1989) and to inhibit the Ca-pump enzyme, Ca^{2+}-ATPase, in cardiac SR (Pitts et al., 1978).

In summary, the arrhythmogenic activity of ouabain and PalCar in cardiomyocytes can be attributed to the appearance of I_{TI}. They are Na-K pump inhibitors, increasing $[Na]_i$, thus facilitating Ca^{2+} influx into the cell through the Na/Ca exchanger. The rising $[Ca]_i$ triggers Ca^{2+} release from the SR, which in turn leads to I_{TI}. The major part of I_{TI} is mediated by $I_{Na/Ca}$ in mammalian ventricular muscle cells (Arlock & Katzung, 1985; Benndorf et al., 1991; Fedida et al., 1987; Kimura, 1988; Lipp & Pott, 1988; Marban & Wier, 1985). However, the Ca^{2+}-activated nonselective cation channel may also contribute.

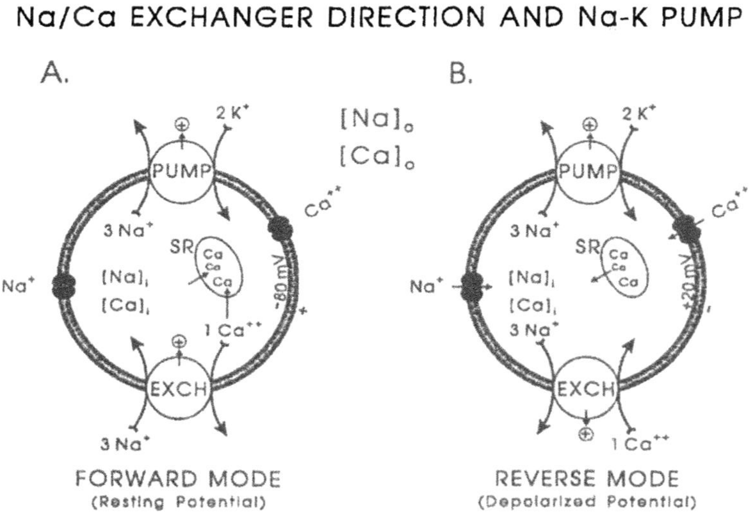

Fig. III-8. Model of a working hypothesis for the functional inter-relationship of the electrogenic Na/Ca exchanger and the electrogenic Na-K pump, as related for a myocardial cell. The Na-K pump normally (3 Na^+_o : 2 K^+_i) produces a net *outward* current which is *hyperpolarizing* (adding a small voltage ($I_p \times R_m$) to E_{diff}). Inhibition of the Na-K pump by ouabain (ouab) or palmatyl-carnitine (PalCar) depresses or abolishes the net pump current, and therefore removes this hyperpolarizing influence. **A:** The Na/Ca exchanger operating in the **forward mode** (3 Na^+_o : 1 Ca^{2+}_i) in a resting cell (-80 mV) produces a net *inward* current which is *depolarizing*, thereby tending to **cancel** out the membrane voltage change produced by the Na-K pump. **B:** The Na/Ca exchanger operating in the **reverse mode** (3 Na^+_i : 1 Ca^{2+}_o) during the plateau of the cardiac AP (+20 mV) produces a net *outward* current which is *hyperpolarizing*, thereby **adding** to the membrane voltage change produced by the Na-K pump. During the AP, the energetics become favorable for reversing of the Na/Ca exchanger (see text). Increased Na^+ entry through the V-dependent Na^+ channels during the AP, if coupled to Na-K pump inhibition, would lead to Na^+ overload, and because of the Na/Ca exchange, would lead to Ca^{2+} overload of the cell and SR. Ca overload causes Ca^{2+} to be released spontaneously from the SR, thus stimulating the forward mode of the exchanger and producing a depolarizing inward current which is the delayed after-depolarization (DAD).

Appendix III

DERIVATION OF Ca/Na EXCHANGE DISTRIBUTION RATIO

This section presents one way to derive the distribution ratios for Ca^{2+} and Na^{+} across a membrane at equilibrium with respect to the Ca/Na exchange reaction in which 2 Na^{+} ions are exchanged for 1 Ca^{2+} ion. We may start with the energy available from the downhill transport of 2 mol of Na^{+}

$$\Delta G_{\mathrm{Na}} = 2 z_{\mathrm{Na}} F \left(E_m - E_{\mathrm{Na}}\right) \qquad \text{eq. A-1}$$

where z_{Na} is the valence of Na^{+}. The energy required for the uphill transport of 1 mol of Ca^{2+} is

$$\Delta G_{\mathrm{Ca}} = z_{\mathrm{Ca}} F \left(E_m - E_{\mathrm{Ca}}\right) \qquad \text{eq. A-2}$$

where z_{Ca} is the valence of Ca^{2+}. At equilibrium,

$$\Delta G_{\mathrm{Na}} = \Delta G_{\mathrm{Ca}} \qquad \text{eq. A-3}$$

and therefore

$$2 z_{\mathrm{Na}} F \left(E_m - E_{\mathrm{Na}}\right) = z_{\mathrm{Ca}} F \left(E_m - E_{\mathrm{Ca}}\right) \qquad \text{eq. A-4}$$

Canceling out F and substituting the valences gives

$$2\,(1)\left(E_m - E_{\mathrm{Na}}\right) = (2)\left(E_m - E_{\mathrm{Ca}}\right) \qquad \text{eq. A-5}$$

$$2E_m - 2E_{\mathrm{Na}} = 2E_m - 2E_{\mathrm{Ca}} \qquad \text{eq. A-5a}$$

Canceling out the 2 E_m terms, rearranging, and dividing each side by 2 gives

$$E_{\mathrm{Ca}} = E_{\mathrm{Na}} \qquad \text{eq. A-5b}$$

Substituting from the Nernst equation (see Chapter V)

$$-30\text{ mV}\log\frac{[\mathrm{Ca}^{2+}]_i}{[\mathrm{Ca}^{2+}]_o} = -60\text{ mV}\log\frac{[\mathrm{Na}^{+}]_i}{[\mathrm{Na}^{+}]_o} \qquad \text{eq. A-6}$$

$$\log\frac{[\mathrm{Ca}^{2+}]_i}{[\mathrm{Ca}^{2+}]_o} = 2\log\frac{[\mathrm{Na}^{+}]_i}{[\mathrm{Na}^{+}]_o} \qquad \text{eq. A-6a}$$

Because two times the log of a number is equal to that number squared, and by canceling the logs, one obtains

$$\frac{[\mathrm{Ca}^{2+}]_i}{[\mathrm{Ca}^{2+}]_o} = \frac{[\mathrm{Na}^{+}]_i^2}{[\mathrm{Na}^{+}]_o^2} \qquad \text{eq. A-7}$$

which is the same as Equation 5 in Chapter III.

Chapter IV

DIFFUSION AND PERMEABILITY

A. Diffusion and Diffusion Coefficient

Any substance in solution (or in a gas) tends to move from regions of higher concentration to regions of lower concentration until the substance is uniformly distributed. (The diffusion equations also apply to the diffusion of heat in solids.) Once uniformly distributed (i.e., at equilibrium), the molecules of the substance continue to move about, but the net movement is zero. In the absence of convection currents, the movement of the molecules is by diffusion. Diffusion occurs because of the **random thermal motion** of the molecules. If a region of high solute concentration is adjacent to one of low solute concentration, separated by an imaginary plane, the probability is that there will be more molecules per unit of time crossing the plane from the side of higher concentration to the side of lower concentration than there are crossing in the opposite direction. Thus, there is a flux or movement of molecules in both directions (unidirectional fluxes), but the net flux is from the side of higher concentration to the side of lower concentration. Now if the imaginary plane were replaced with a thin membrane permeable to the molecules, then the same situation applies; namely, the particles diffuse from the side of higher concentration to the side of lower concentration across the membrane. Because in a living cell it is often assumed that the inside and outside solutions are relatively well stirred, and because diffusion of most substances through the cell membrane is much slower than through a free solution, our concern is about the diffusion of substances through the cell membrane, i.e., the rate-limiting step. We assume for simplicity that the solutions on either side are well stirred, i.e., that there are no **concentration gradients** within the bulk solution on either side of the membrane (although there probably are **unstirred layers** near the membrane). We confine ourselves here to the diffusion of small molecules or ions across membranes.

Fick's diffusion equation states that the rate of transfer of uncharged molecules across a membrane is equal to the concentration gradient (*dC/dx*) times the area of the membrane (A_m) times a **coefficient of diffusion** (*D*)

$$\frac{dQ}{dt} = -DA_m \frac{dC}{dx} \qquad \text{eq. 1}$$

where Q is the amount of substance (in mol) transported, t is the time, C is the molecular concentration, and $\boldsymbol{x}$ is the distance. (Although concentration is used throughout this chapter for clarity, activity is actually meant, i.e., the thermodynamically active concentration, which is the concentration times the **activity coefficient**.) For a thin membrane, when the solution is well stirred on either side, the concentration gradient in the steady state is equal to the difference in the molecular concentration (ΔC) of the solute on both sides of the membrane divided by the thickness of the membrane (Δx)

$$J' = -DA_m \frac{\Delta C}{\Delta x} \qquad \text{eq. 2}$$

where $dQ/dt = J'$, where J' is the rate of movement (or flux) of the substance across the membrane in mol/sec. Now we can transpose A_m from the right side of the equation to the left side, and letting $J'/A_m = J$, where J is the flux in mol/sec·cm^2

$$J = -D \frac{\Delta C}{\Delta x} \qquad \text{eq. 3}$$

Thus, the **net flux** is equal to the concentration gradient times the **diffusion coefficient**. The diffusion coefficient is a constant for a given substance and membrane under a given set of conditions, and the coefficient is given in cm^2/sec

$$\frac{\text{mol}}{\text{sec} \cdot \text{cm}^2} = -\frac{\text{cm}^2}{\text{sec}} \frac{\text{mol}/\text{cm}^3}{\text{cm}} \qquad \text{eq. 3a}$$

The negative sign in Equation 3a refers to the direction of the net flow, namely, down the concentration gradient.

The diffusion coefficient of substances in a free solution is dependent on the molecular size (and shape, for large molecules). For ions, the smaller the **unhydrated radius**, the greater the charge density is, which means that more water molecules are held in the hydration shells (larger **hydrated radius**); the larger water shell causes diffusion to be slower. The hydrated and unhydrated radii of some relevant ions are given in Table IV-1.

The time required for diffusion to become 50%, 63%, or 90% complete varies directly with the square of the distance, and inversely with the diffusion coefficient. Thus, diffusion is extremely fast over short distances (e.g., 10–1000 nm) but exceedingly slow over long distances (e.g., 1 cm). For a small particle, like K^+ or acetylcholine (ACh^+), with a D value of about 1×10^{-5} cm^2/sec, the time required for 90% equilibration to be reached for a distance of 1 μm is about 1 msec. Table IV-2 shows how the time changes throughout a wide range of values as a function of distance. The chef knows that the time required for a meat roast to cook in the middle is highly

TABLE IV-1

Physicochemical properties of selected ions.

Ion	Limiting equivalent conductivity (cm^2/mho-eq)		Activity coefficient	Crystallo-graphic radius	Singly hydrated radius	Hydrated volume	Heat of hydration	No. of water molecules		Relative shielding
	25° C	35° C		Å	Å	$Å^3$	(keal g ion^{-1})			
Na^+	50.10	61.5	0.778	0.96	3.67	150	-115	5	(4-6)	0.432
K^+	73.50	88.2	0.770	1.33	4.05	(150)	-90	(5)	(3-6)	0.224
Cl^-	76.35	92.2	--	1.81	3.92	90	-59	(3)	(0-6)	(0.073)
Mg^{++}	53.05		0.528	0.65	3.60	360	-501	12		2.18
Ca^{++}	59.50		0.518	0.99	3.70	310	-428	10		0.812
Sr^{++}	59.45		0.515	1.13	3.85	310	-381	10		0.624
Ba^{++}	63.63		0.508	1.35	4.08	290	-347	9-10		0.415

Values taken from Robinson, R.A., and Stokes, R.H., Electrolyte Solutions, London, Butterworth & Co. (Publishers), Ltd., 1959, and from Bockris, J. O'M., and Conway, B.E., Modern Aspects of Electrochemistry, London, Butterworth & Co. (Publishers), Ltd., 1956.
Values taken for singly-hydrated radius taken from Mullins (1961).

Activity coefficient for each cation given for the Cl^- salt at 0.1 M and at 25° C.

Singly hydrated radius equals crystal radius + 2.72 Å (diameter of water molecule) for cations with crystal radii between 0.9 to 1.7 Å and crystal radius + 2.23 Å for anions.

Values in parentheses are estimated values.

Relative shielding was calculated from the number of water molecules per A^2 of unhydrated surface area. The order of relative shielding, from least to most shielded, is: Cl < K < Na ≅ Ba < Sr < Ca < Mg. This order is similar to that for the heats of hydration: Cl < K < Na << Ba < Sr < Ca < Mg. Thus, Cl^- sheds its hydration the easiest and Mg^{++} the hardest. The cations which are least shielded and which can shed their hydration the easiest should get through the cell membrane most easily if pore diameter is the limiting factor. If so, P_K should be relatively large, and Mg^{++} should be least active electrophysiologically.

TABLE IV-2

Calculations of time required for diffusion to become 90% complete.

Distance		Time	
100	Å	0.1	µsec
0.1	µm	0.01	msec
1	µm	1	msec
10	µm	100	msec
100	µm	10	sec
1	mm	16.7	min
1	cm	28	hr

Diffusion time varies with square of diffusion distance. These values are for a substance having a diffusion coefficient (D) of about 1 x 10^{-5} cm^2/sec in free solution. The diffusion time is inversely proportional to D. Einstein's approximation equation is:

$$\bar{d} = \sqrt{2Dt}$$

where $\bar{d}$ is the mean displacement of the diffusing particles along the x axis at any time (t). Thus, the mean time (T) it takes for a particle to diffuse to a given distance (d) is given by:

$$T = \frac{d^2}{2D}$$

Calculation from this equation gives a mean time of 0.5 msec for a diffusion distance of 1.0 µm and a diffusion coefficient of 1 x 10^{-5} cm^2/sec.

dependent on the thickness of the roast, and the physiologist knows that there is a critical thickness (e.g., 0.5–1.0 mm) of a muscle bundle in a bath that allows adequate diffusion of oxygen to the cells within the core of the strip, no matter how vigorously the bath is oxygenated.

The diffusion coefficient across cell membranes for various substances is generally greater when the molecular size of the substance is small and when the lipid solubility is high; that is, small molecules of high lipid solubility (i.e., less polar and nonpolar molecules) penetrate through the membrane the fastest. Most nonpolar molecules pass through the lipid bilayer matrix of the membrane, i.e., not through special sites (e.g., water-filled pores or channels). Small charged ions (e.g., Na^+, K^+, Cl^-, Ca^{2+}) appear to pass through water-filled channels, some of which have a voltage-dependent gating mechanism; these channels can exhibit a high degree of selectivity for specific ions, the selectivity orders not being solely based on the hydrated or unhydrated sizes of the ions.

As discussed in Chapter I, one type of ionophore (antibiotic) can act as a cage carrier for K^+, the K^+- ionophore complex being lipid soluble and presumably passing through the lipid bilayer matrix. Other types of carrier molecules are normally present in the cell membrane for transport of substances such as glucose and amino acids. Some of these carriers are for downhill transport only, whereas others can provide uphill transport against electrochemical gradients. One type of carrier-mediated transport (e.g., for glucose in heart cells) is known as facilitated diffusion (in contrast with the so-called simple diffusion described in the preceding paragraphs); that is, the downhill movement of the substance across the membrane is enhanced by the presence of the carrier. Another type of carrier-mediated transport is known as **exchange diffusion**. In this type of diffusion, one molecule of substance (or ion) inside the cell is exchanged for one molecule outside the cell; i.e., there is no net movement.

As discussed in Chapter II, the Na-K exchange pump sometimes can be made to act as a Na/Na exchange diffusion carrier. In exchange diffusion the carrier cannot move across the membrane in either direction unless loaded with the substance. One characteristic of carrier-mediated transport systems is that they all exhibit saturation kinetics; i.e., the rate of transport increases with an increase in the substrate concentration up to some maximum, after which transport levels off, presumably due to a finite number of available carriers. Other characteristics of carrier-mediated transport systems include: competitive inhibition (two or more substances may compete for binding to the carrier molecule), noncompetitive inhibition (a nontransported substance may bind to the carrier and prevent the carrier from combining with its usual substrate), specificity (carrier is more or less specific for one physiological substance, different substances having different carriers), and a high temperature coefficient (Q_{10}).

B. Permeability Coefficient

Returning to Equation 3, we can make one final combination. Because the cell membrane is relatively fixed in thickness, we can combine Δx and D into a new coefficient: $P = D/\Delta x$. Substitution into Equation 3 gives

$$J = -P\Delta C = -P\,(C_o - C_i) \qquad \text{eq. 4}$$

where P is the **permeability coefficient**, C_i is the internal concentration, and C_o is the external concentration. The units of the permeability coefficient are cm/sec, i. e., that of a velocity

$$\frac{\text{mol}}{\text{sec} \cdot \text{cm}^2} = -\frac{\text{cm}}{\text{sec}} \frac{\text{mol}}{\text{cm}^3} \qquad \text{eq. 4a}$$

Hence, the net flux of a nonelectrolyte across a membrane is equal to the permeability coefficient times the difference in concentration across the membrane. Values for the permeability coefficient of K^+ (P_K) across resting striated muscle membrane are 1×10^{-6} cm/sec. Some P values are given in Table IV-3 for some selected heart tissues. Note that although diffusion coefficients apply to diffusion in a free solution or in membranes, permeability coefficients only apply to membranes. The higher the diffusion coefficient for movement of a substance across a membrane, the higher the permeability coefficient is. The unidirectional fluxes are given by

$$J_i = -PC_o \qquad \text{eq. 5}$$

and

$$J_o = +PC_i \qquad \text{eq. 6}$$

J_i is the influx of the substance and J_o is the outflux (efflux). Thus

$$J = J_i - J_o \qquad \text{eq. 7}$$

$$= -PC_o - (+PC_i) \qquad \text{eq. 7a}$$

$$J = -P(C_o - C_i)$$

Equations 5 and 6 show that the influx of a substance is equal to its P value times the external concentration, whereas the efflux is equal to the P value times the internal concentration. Conversely, the permeability is equal to the ratio of flux to concentration: influx to C_o or efflux to C_i.

TABLE IV-3

Summary of internal K^+ and Na^+ concentrations and activities for some selected heart tissues.

Preparation	Resting Potential mV	P_K cm/s ($\times 10^{-7}$)	P_{Na} cm/s ($\times 10^{-7}$)	$\frac{P_{Na}}{P_K}$	$[K]_i$ mM	$[Na]_i$ mM	Reference
Chick embryonic, 19 days old		3.10	0.053	0.017	122	15	Carmeliet et al., 1976
Rabbit papillary	-77.5	--	--	--	135 82.6#	32.7 5.7#	Lee & Fozzard, 1975
Rabbit papillary	-86	--	--	--	119##	--	Akiyama & Fozzard, 1975
Rabbit ventricular	-76	--	--	--	83.1#	--	Fozzard & Lee, 1976
Cow Purkinje	-75	1.66	--	--	--	--	Carmeliet & Verdonck, 1977
Sheep Purkinje	--	--	--	--	160	--	Carmeliet & Bosteels, 1969

#Measured with ion-selective microelectrodes. ##Calculated from the internal K^+ concentration and activity coefficient. Unless otherwise specified, all ion concentrations are based on total tissue analyses, and all permeability coefficients are for tissues bathed in normal Ringer's solution.

Equations 4–6 only apply to uncharged molecules. If there is a net charge on the molecule, then the unidirectional and net fluxes are also determined by any electrical field that may exist across the membrane. The equation for the net flux (obtained from the constant field assumption or other suitable electrochemical flux theory) is in the general form

$$J = P f(E_m) \left[C_o - C_i e^{E_m F/RT} \right] \qquad \text{eq. 8}$$

where $f(E_m)$ is some function of the potential difference across the membrane, and F, R, and T are the Faraday constant, the gas constant, and absolute temperature, respectively. Note the similarity of Equation 8 to Equation 4, except for the membrane potential terms. When $f(E_m)$ is equal to $E_m F/RT$, then this term is dimensionless (EF is the electrical energy, whereas RT is the thermal energy): $F/RT = (0.026\ \mathrm{V})^{-1}$ at 37° C. Similarly, the term exp $E_m F/RT$ is dimensionless; therefore, the units are the same as those given in Equation 4a. Equation 8 is further pursued in the Appendix A to Chapter VI.

The permeability coefficient (for a charged ion) can be better understood by considering the relationship (in this relationship P is based on the solution concentration and not on the membrane concentration)

$$P = U \frac{\beta RT}{\delta F} \qquad \text{eq. 9}$$

$$\frac{\mathrm{cm}}{\mathrm{sec}} = \frac{\mathrm{cm/sec}}{\mathrm{V/cm}} \frac{1}{\mathrm{cm}} \frac{\mathrm{V}}{1}$$

where U is the mobility of the ion through the membrane (units of a velocity per unit driving - electrophoresing - voltage gradient, cm/sec per V/cm); β is the partition coefficient for the ion between the bulk solution and the edge of the membrane, and is dimensionless; δ is the membrane thickness, in cm; and $RT/F = 0.026$ V at 37° C. In essence, Equation 9 indicates that the permeability coefficient is a direct function of the mobility of the ion through the membrane; in other words, it is a velocity in a unit electrical field.

The discussion about Equations 5–7 tells us that the unidirectional fluxes of an ion should be proportional to the electrochemical potentials ($\overline{\mu}$) for that ion; that is, the influx should be a direct function of the electrochemical potential of the ion outside ($\overline{\mu}_o$), and the efflux should be a direct function of the electrochemical potential inside ($\overline{\mu}_i$)

$$J_i \propto \overline{\mu}_o \qquad \text{eq. 10}$$

$$J_o \propto \overline{\mu}_i \qquad \text{eq. 11}$$

where $\overline{\mu}_o$ is a function of the outside concentration and the absolute electrical potential (Ψ_o) and $\overline{\mu}_i$ is a function of the inside concentration and electrical potential (Ψ_i). The potential difference across the membrane (E_m) equals $\Psi_i - \Psi_o$. The net flux is from the side of greater electrochemical potential to the side of lesser electrochemical potential.

The electrochemical potential ($\overline{\mu}$) is a measure of the useful energy, and its units are in joul/mol (just as voltage is in joul/coul). Electrochemical potential is composed of a chemical part (μ_c) and an electrical part (μ_e)

$$\overline{\mu} = \mu_c + \mu_e \qquad \text{eq. 12}$$

The chemical part is equal to

$$\mu_c = \mu_c^{\,0} + RT \ln a \qquad \text{eq. 13}$$

where μ_c^o is the standard **electrochemical potential** at standard temperature and pressure, and a is the activity (activity coefficient, γ, times the concentration). In animal cells, however, there is neither a significant temperature difference nor a substantial pressure difference across the cell membrane. Therefore, the maximal electrochemical potential for Na^+ inside is given by

$$\overline{\mu}_i^{\,\mathrm{Na}} = \mu_{\mathrm{Na}}^{\,0} + RT \ln(\gamma_i[\mathrm{Na}]_i) + zF\Psi_i \qquad \text{eq. 14}$$

where ψ_i is the internal potential, and $zF\Psi_i$ is the electrical energy per mole. The difference in electrochemical potential between inside and outside ($\Delta\overline{\mu} = \overline{\mu}_i - \overline{\mu}_o$) then is

$$\Delta\overline{\mu}^{\,\mathrm{Na}} = RT \ln \frac{[\mathrm{Na}]_i}{[\mathrm{Na}]_o} + zF\Delta\Psi \qquad \text{eq. 15}$$

where $\Delta\Psi = \Psi_i - \Psi_o$, and is the same as E_m. The activity coefficients cancel out, assuming $\gamma_i = \gamma_o$.

The net passive fluxes for each cation must be equal and opposite to a net active flux against the electrochemical gradient under steady-state conditions. For example, the net influx of Na^+ down its large electrochemical gradient ($E_m - E_{Na}$) must be balanced by an active pumping of Na^+ out of the cell; otherwise the cell continues to gain Na^+. Similarly, the net efflux of K^+ down its small electrochemical gradient ($E_m - E_K$) must be balanced by an active pumping of K^+ into the cell; otherwise the cell loses K^+.

The electrical current (i, in A) is equal to the flux (J', in mol/sec) times zF

$$i = J'zF \qquad \text{eq. 16}$$

or

$$I = JzF \qquad \text{eq. 17}$$

$$\frac{\text{amp}}{\text{cm}^2} = \frac{\text{mol}}{\text{sec}\cdot\text{cm}^2}\,\frac{\text{coul}}{\text{mol}}$$

where I is the current density in amp/cm^2 , and J is the flux density in mol/sec·cm^2.

C. Ussing Flux Ratio Equation

As pointed out in the preceding paragraphs, the absolute unidirectional fluxes are determined also by the permeability coefficient. Ussing (1949) developed the so-called **flux ratio equation**, in which the ratio of influx to efflux is used (permeability cancels out). Specifically, for the ratio of Na^+ fluxes (by simple electrodiffusion), the following applies

$$\frac{J_i^{\text{Na}}}{J_o^{\text{Na}}} = \frac{[\text{Na}^+]_o}{[\text{Na}^+]_i}\, e^{-E_m F/RT} \qquad \text{eq. 18}$$

Thus, for Na^+ the ratio of influx: efflux (passive) in a resting membrane (assuming a resting potential of -80 mV) is

$$\frac{J_i^{\text{Na}}}{J_o^{\text{Na}}} = \frac{150\text{ mM}}{15\text{ mM}}\, e^{+80\text{ mV}/26\text{ mV}} \qquad \text{eq. 18a}$$

$$= 10 \times e^{3.08}$$

$$= 217$$

Thus, the passive influx of Na^+ should be 217 times greater than the passive efflux, because of the large electrochemical gradient directed inward; if the free $[Na]_i$ were actually only 1.5 mM, then the flux ratio would be 2,150. The flux ratio for K^+ would be

$$\frac{J_i^{K}}{J_o^{K}} = \frac{4\text{ mM}}{150\text{ mM}} e^{+80\text{ mV}/26\text{ mV}} \quad \text{eq. 18b}$$

$$= \frac{1}{37.5} e^{3.07}$$

$$= 0.574$$

Thus, the passive influx should be 0.57 times the passive efflux. The K^+ diffusion potential (E_K) is only slightly greater (more negative by about 14 mV) than the resting potential, and so the passive flux ratio should be close to one.

A modified form of Equation 18 can be obtained by substituting the ratio of ions with $e^{E_i F/RT}$ derived from the Nernst equation and reducing

$$\frac{J_i^{i}}{J_o^{i}} = e^{-(E_m - E_i)F/RT} \quad \text{eq. 19}$$

where E_i is the equilibrium potential for the cation in question, and J_i^i and J_o^i are the inward and outward fluxes, respectively, for ion *i*. [An alternate derivation of Equation 19 was given by Hodgkin and Huxley (1952b); however, they used the early **sign convention** of referring all potentials (E_m and E_i) of outside solution to inside solution, rather than the newer convention of **inside solution to outside solution**.] Thus for Na^+:

$$\frac{J_i^{\text{Na}}}{J_o^{\text{Na}}} = e^{-[-80\text{ mV}-(+60\text{ mV})]/26\text{ mV}} \qquad \text{eq. 19a}$$

$$= e^{+5.38}$$

$$= 217$$

This value obtained for the flux ratio is identical to that obtained from Equation 18. One advantage of Equation 19 is that it is obvious at a glance that when $E_i = E_m$, the flux ratio is exactly 1.0, because $e^0 = 1$, where e is the standard base of the natural logarithm. Thus, if Cl^- is passively distributed so that $E_{Cl} = E_m$, its flux ratio should be 1.0; i.e., influx equals efflux, so there is no net flux. The larger the difference between E_i and E_m, i.e., the farther the ion is off equilibrium, the greater the flux ratio.

Sometimes Equations 18 and 19 do not exactly fit the experimental facts. The data are better fitted if the exponential term contains another factor (N)

$$\frac{J_i^i}{J_o^i} = e^{-(E_m - E_i)\left(\frac{F}{RT}\right)N} \qquad \text{eq. 20}$$

The best fit of the data is when N has a value of 2.5 to 4.0, depending on the membrane under investigation. The interpretation given to N is that if the length of the water-filled pore that the ion must traverse across the membrane is much longer than the ion diameter, as is likely, then for so-called **single-file diffusion**, N hits from the same side are required for the ion to complete its journey across the membrane. One could consider, for example, that there are 2, 3, or 4 **potential energy wells**, or chain of reactive sites, along the length of the pore, and the only way for the ion to escape the well is to get a **kinetic bump** from an adjacent ion in the file. Complete permeation of an ion through the pore is **more likely to happen if the ion is moving in the same direction as the majority**, i.e., down the electrochemical gradient. Therefore, this factor makes the flux ratio much greater than would otherwise be predicted.

Appendix IV

DIFFUSION

A. Gases

The **diffusion coefficients (*D*)** of different gases are inversely proportional to the square roots of their masses; see the discussion in Davson's textbook (1964)

$$DM^{1/2} = \text{constant} \qquad \text{eq. A-1}$$

In gases, the molecules have random thermal motion (i.e., **Brownian movement**). The molecules are free to move, being subject to only small intermolecular forces, because the molecules are relatively far apart. The kinetic energy of the gas molecules is given by $1/2\ NM\ \theta^2$, where N is the number of molecules, M is their molecular weight, and θ is their mean velocity. Because $DM^{1/2}$ is constant for different gases, the average energy of their molecules is the same at the same temperature and pressure. Hence, the average velocities of the molecules are inversely proportional to the square root of their masses. Because the velocity of the molecules determines the speed of diffusion, the diffusion coefficient is inversely proportional to the square root of the molecular weight

$$D \propto \frac{1}{M^{1/2}} \qquad \text{eq. A-1a}$$

B. Liquids

In liquids, on the other hand, the molecules are close together and within each others' spheres of mutual attraction. Hence, for any molecule to move, it must break away from its surrounding molecules. Diffusion in liquids consequently occurs in a **discontinuous** manner, a molecule being able to move only when it has acquired sufficient energy by collision to break away from its neighbors and push aside other molecules. Thus, the **molecule diffuses in a series of jumps**, each jump requiring a certain critical **activation energy**. When the attractive forces between molecules are weak, as in a homopolar solvent, diffusion is rapid, whereas with substances like glycerol diffusing in water, the attractive forces between the OH groups of glycerol and water are strong due to formation of hydrogen bonds (the same is true for HOH molecules diffusing in water), and the diffusion coefficients are low. Because of the activation energy required to make each jump possible, in liquids the diffusion coefficient of a substance is inversely proportional to the square root of its molecular weight times the **temperature coefficient (Q_{10})** raised to a power, according to Danielli (1943) (see Davson, 1964)

$$DM^{1/2}Q_{10}^{(T+10)/10} = \text{constant} \qquad \text{eq. A-2}$$

or

$$D \propto \frac{1}{M^{1/2}Q_{10}^{(T+10)/10}} \qquad \text{eq. A-2a}$$

This relationship, derived from kinetic theory, demonstrates that the Q_{10} for diffusion of a substance in water depends on the activation energy necessary for a jump. If a large amount of energy is required, only a few molecules have the necessary energy to diffuse at any moment; raising the temperature increases the number of molecules with the required energy and thereby has an appreciable effect on the net rate of diffusion. If only a small amount of activation energy is required, a greater fraction of the molecules possesses this minimal energy at any moment, and so raising the temperature has a lesser effect on the net rate of diffusion. Elevation of temperature has two effects: an increase in the speed of the molecules and an increase in their energy. The increase in speed causes an increase in the rate of diffusion; the effects of increased energy depend on how much energy is required.

To repeat, the more the activation energy, the greater the effect of a rise in temperature, and thus the greater the value of Q_{10}. When Q_{10} is close to 1.0, then Equations A-2 and A-2a reduce to Equations A-1 and A-la. The Q_{10} for diffusion of Na^+ or K^+ in water is only about 1.22. In this case, the mass of the hydrated ion, rather than the atomic weight, must be used. Although Na^+ has a lower atomic weight than K^+, Na^+ attracts and holds a larger hydration shell because of its greater charge density (same charge as K^+, but smaller unhydrated ion size), and therefore Na^+ diffuses in water considerably slower than K^+ (ratio of about 1 : 2).

C. Membranes

The cell membrane constitutes a barrier to diffusion. In penetrating through the membrane (through the lipid bilayer matrix), a noncharged small solute molecule must (a) detach itself from its surrounding solvent molecules and jump into the membrane phase; (b) move through the thickness of the membrane, perhaps by a series of small jumps; and (c) detach itself from the membrane environment and jump into the solvent phase again on the opposite side of the membrane. The permeability of the membrane to the solute depends greatly on the activation energy required by the molecule to jump into and out of the membrane. If the activation energy is high, the permeability is slow and the Q_{10} is high; if the activation energy is low, the permeability is great and the Q_{10} is low. With substances containing polar groups, such as -OH, -NH_2, -SO_3H, the amount of energy necessary for detachment from the surrounding water molecules is great because of the hydrogen bonds formed. Thus, the rate of permeation of a water-soluble substance is mainly determined by the substance's ease of passing from water to lipid; its **oil-water partition coefficient** is a measure of this ease. Equations A-2 and A-2a also apply if the permeability coefficient (*P*) is substituted by the diffusion coefficient (*D*) of the solute in the membrane.

Chapter V

CONCENTRATION CELLS AND EQUILIBRIUM POTENTIALS

Given the ion distributions inside and outside the cell, distributions that are maintained under steady-state conditions, we must now examine how such ion concentration gradients give rise to **diffusion potentials**. To do this, let us start with a **concentration cell** that we can construct in the laboratory, as diagrammed in Figure V-1. The chamber is completely divided into two compartments by a thin membrane made of celloidin, which has negatively charged pores. A solution of any salt at one concentration is placed in one compartment (side 1), and a solution of the same salt but at another concentration is placed in the other compartment (side 2). In the example shown, 0.1 M NaCl is in side 1 and 0.01 M NaCl is in side 2. There is a tendency for Na^+ and Cl^- ions to diffuse down their **concentration gradients** from side 1 to side 2. The force for this diffusion process is the **random motion** of the particles, motion due to **thermal energy**. The probability is that the side of higher concentration will have more particles moving across the membrane from side 1 to side 2 per unit time than the side of lower concentration has moving across the membrane in the opposite direction.

The negatively charged Cl^- ions, however, cannot pass across the membrane, because of the negatively charged pores (**like charges repel**), as shown in the lower portion of Figure V-1. Therefore, the net diffusion of Na^+ from side 1 to side 2 brings a surplus of positive charges to side 2 and leaves a deficiency of positive charges, or excess of negative charges, on side 1. In this manner, a potential difference (PD) is developed across the membrane, the side of higher

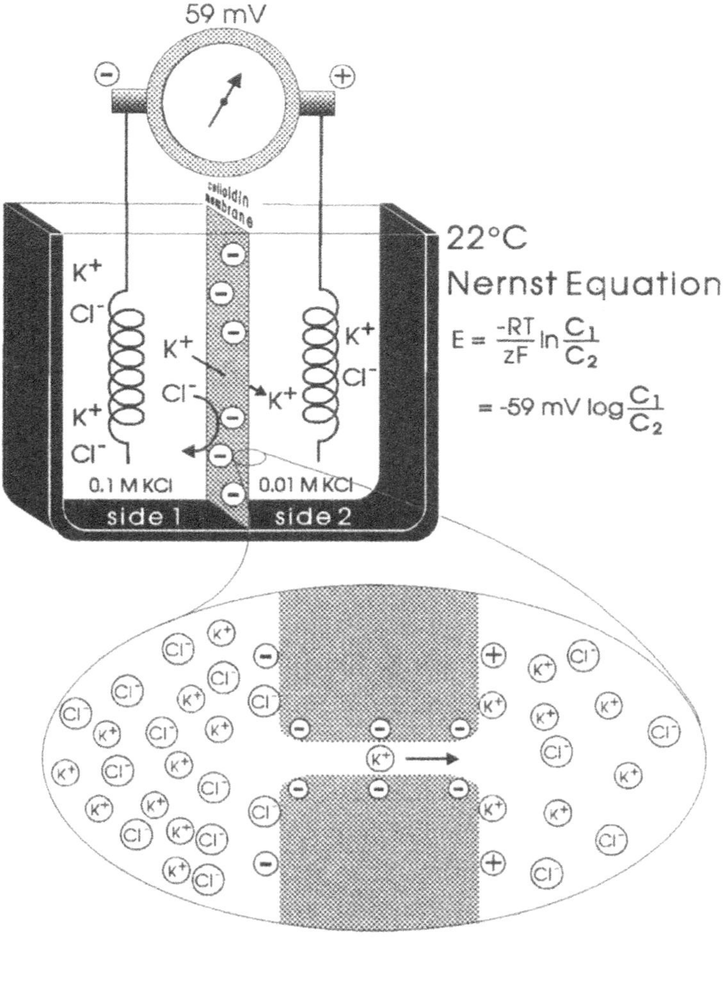

FIGURE V-1. **Upper diagram:** Concentration cell diffusion potential developed across artificial membrane containing negatively charged pores. Thus the membrane is impermeable to Cl^- ions but permeable to cations such as Na^+. Concentration gradient for Na^+ causes a potential to be generated, the side of higher Na^+ concentration becoming negative. **Lower diagram:** Expanded diagram of a water-filled pore in the membrane, showing the permeability to Na^+ ions but lack of penetration of Cl^- ions. Potential difference is generated by charge separation, a slight excess of Na^+ ions being held close to the right-hand surface of the membrane; a slight excess of Cl^- ions is plastered up close to the left surface.

concentration becoming negative when the membrane is permeable only to positive ions. The PD can be recorded by an appropriate voltage-recording device with an electrode in each compartment solution.

The system **equilibrates** within a few seconds, the PD reaching a maximum of 60 mV (at 30° C). The magnitude of this Na^+ **equilibrium potential**, symbolized as E_{Na}, can be calculated from the **Nernst equation.*** The Nernst equation holds true only if there is **no net current flow**; hence although equilibrium potential and diffusion potential are often used interchangeably, strictly speaking it is an equilibrium potential. The Nernst equation written in general form is

$$E_i = -\frac{RT}{zF} \ln \frac{C_1}{C_2} \qquad \text{eq. 1}$$

where E_i is the **equilibrium potential** for any ion, R is the **gas constant** (8.3 joul/K per mol), T is the **absolute temperature** (°K = °C + 273°), z is the **valence** of the ion, F is the **Faraday constant** (96,500 coul/equivalent), and C_1 and C_2 are the concentrations of the ion on side 1 and side 2. The Faraday constant is derived from the number of ions per mol, **Avogadro's number** of 6.02×10^{23}, times the charge per univalent ion, 1.60×10^{-19} coul (see Appendix to this chapter). The Nernst equation can be derived from thermodynamic considerations (see Appendix to this chapter). The direction of

---

* The Nernst equation was devised by Walther Nernst (1864-1941) who was born in Prussia, received a doctorate in physics in 1887, and became a physical chemist while working with Ostwald. He is best known for his development of the third law of thermodynamics and for his contributions to electrochemistry. He received the Nobel prize in chemistry in 1920.

the battery, but not its magnitude, is determined by the sign; a negative sign in front of the log of a ratio merely inverts the ratio. Because the side of higher concentration is always negative for positive ions and positive for negative ions, the direction of the battery can also be easily determined. By lumping the constants and the factor of 2.303 to convert from natural logarithms to logarithms to the base 10, the Nernst equation becomes

$$E_i = \frac{-0.061\ \mathrm{V}}{z} \log \frac{C_1}{C_2} \qquad \text{eq. 2}$$

For univalent ions, the factor is 0.059 V at 25° C and 0.061 V at 35° C; for divalent ions, the factor is 0.030 at 30° C. Thus, for a concentration ratio of 10, E_{Na} is 60 mV at 30° C.

The Na^+ does not continue to diffuse from side 1 to side 2 until the two sides are at equal concentrations, because the PD that is rapidly built up tends to oppose further net diffusion. Because side 2 becomes positively charged to a large voltage, this charge opposes further net diffusion of positively charged Na^+ ions (**like charges repel**). Thus the electrical gradient — built up because of the ionic gradient — in turn acts in opposition to the concentration gradient. The Nernst equation gives the PD necessary to be built up to exactly oppose the concentration gradient (this accounts for the negative sign in the equation). The concentration (or chemical) force is expressed in electrical terms. At equilibrium, the concentration of Na^+ and Cl^- in both compartments is almost unchanged from the initial values, the change being too small to determine chemically. In other words, only a relatively small separation of charge (Q) is necessary to build up a large PD. An estimation of the charge separation required to produce a PD of 60 mV, assuming a membrane capacitance (C_m) of 1 μF/cm^2 may be obtained from the following relationship

$$Q = C_m V \qquad \text{eq. 3}$$
$$= \left(1 \times 10^{-6} \text{ F/cm}^2\right)\left(60 \times 10^{-3} \text{ V}\right)$$
$$= 60 \times 10^{-9} \text{ coul/cm}^2$$

where V is the voltage. Dividing this value by the Faraday constant (approximately 1×10^5 coul/mol) gives 0.60×10^{-12} mol/cm^2. Thus an excess of only 0.60 pmol Na^+/cm^2 of membrane on side 2, and a concomitant excess of the same amount of Cl^- on side 1, gives rise to the large PD of 60 mV. If the membrane capacitance is lower, then the charge separation required for a given PD is even less.

Thus, only tiny amounts of Na^+ and Cl^- ions, compared to the total amount of Na^+ and Cl^- present in the two solutions, need to be separated to produce the large PD. The charges so separated are plastered up close to the membrane, and electrostatic interactions across the membrane dielectric hold the charges in position. In the two bulk solutions proper (away from the membrane) the numbers of Na^+ (cations) and Cl^- (anions) ions must be exactly equal to one another; i.e., the **principle of electroneutrality** in solution is not violated. Suitable electrodes placed in the two solutions see the charge on each surface of the membrane by means of the low-resistance solutions, and hence the electrodes record a PD; this arrangement effectively causes the two solutions themselves to be at different potentials. Note that the transmembrane resting potential recorded by an intracellular microelectrode is not a function of reversible half-cell potentials, because the solutions bathing the half-cells stay the same whether the microelectrode is inside or outside the cell (see the Appendix to this chapter).

In the preceding discussion of Figure V-1, it was better for didactic purposes to allow only one ion to be able to penetrate across the membrane and to allow the other to be completely nonpenetrating, so that a large PD could be developed. This need not be true, however; that is, whenever one or more ions are at different concentrations on two sides of a membrane, there is a diffusion potential for each ion so distributed, regardless of their permeabilities. Practically, all ions have some finite permeability across the membrane, and therefore a bona fide diffusion potential for each ion can be developed. Thus, the equilibrium potential for Cl^- (E_{Cl}) in the example of Figure V-1 is numerically equal to that for Na^+, but of the opposite sign; that is, the equilibrium potential, represented schematically as a battery in the equivalent circuit for the cell membrane, always is such that the side of higher concentration is positive for an anion and negative for a cation. When the concentrations of a given ion on both sides of the membrane are equal, then the equilibrium potential calculated from the Nernst equation is zero (log $1 = 0$). For a living muscle cell then, there are diffusion potentials for Na^+, K^+, Ca^{2+}, Cl^-, H^+, as well as for other charged molecules or ions, because these ions exist at different concentrations on both sides of the cell membrane. Because for the determination of the resting potential (as well as of the action potentials and postsynaptic potentials) the major ions of importance are Na^+, K^+, Ca^{2+}, and Cl^-, we are only concerned with the equilibrium potentials of Na^+, K^+, Ca^{2+}, and Cl^- (E_{Na}, E_K, E_{Ca}, and E_{Cl}) in the electrical equivalent circuit. It is only these ions that have high enough concentrations, high enough diffusion potentials, and high enough membrane conductances to be able to significantly influence the membrane potential under normal conditions.

An approximation of the internal ion concentrations in cardiac muscle was given in Chapter II (Table II-1), and some more accurate values for $[K]_i$ and $[Na]_i$ were given in Chapter IV (Table IV-3). Table V-1 summarizes the data that Akiyama

TABLE V-1

Concentrations and activities of intracellular K^+ and Na^+ ions in rabbit papillary muscle.

Temperature °C	$[K]_i$ mM	α^i_K mM	γ_K	$[Na]_i$ mM	α^i_{Na} mM	γ_{Na} apparent
35	136	83.4	0.613	32.7	6.4	0.205
25	135	82.6	0.612	32.7	5.7	0.175
36	160	119	0.742	—	—	—

α^i_K: Internal K^+ activity. γ_K: Activity coefficient of K^+. α^i_{Na}: Internal Na^+ activity. γ_{Na}: Activity coefficient of Na^+. The concentrations were determined by flame photometry of entire tissues, and the internal activities were measured by cation-selective glass microelectrodes. The $\alpha_K{}^i$ was considerably lower than the concentration, giving an apparent γ_K of about 0.61; this value is substantially lower than the K^+ activity coefficient in Ringer's solution (0.745). The much lower α^i_{Na}, which causes an apparent γ_{Na} of only about 0.20, is presumably due to compartmentalization of much of the internal Na^+ (e.g., in SR). Data from Lee and Fozzard, 1975.

$[K]_i$ (in isotonic solution) was measured by extrapolation of the resting membrane potential vs. log $[K]_o$ curve to zero potential, and α^i_K was calculated assuming an internal γ_K of 0.742. Data from Akiyama and Fozzard, 1975.

and Fozzard (1975) and Lee and Fozzard (1975) obtained from studies about the activities of the internal K^+ and Na^+, as measured by intracellular cation-selective glass microelectrodes. As can be seen in the table, the internal activity coefficient for K^+ was 0.61 in one study and 0.74 in another; the latter value is virtually identical to that for K^+ in free solution (Ringer's). In contrast, the internal Na^+ activity was much less than the measured Na^+ concentration; this lower activity suggests that most of the internal Na^+ is compartmentalized, presumably in the sarcoplasmic reticulum (SR).

Comparable data for internal Cl^- in heart tissues are given in Table V-2. Measurements of internal Cl^- activity in sheep Purkinje and frog myocardial cells with Cl^--selective microelectrodes gave values of about 14-17 mM. The calculated E_{Cl} values then are substantially lower than the average membrane potential (Table V-2) and thus may be consistent with the view that Cl^- is out of equilibrium; that is, there could be a Cl^- pump directed inward, giving a higher $[Cl]_i$ than what would be predicted on the basis of a passive distribution.

TABLE V-2

Summary of internal Cl^- ion concentration and activity for some selected heart tissues.

Preparation	Resting E_m mV	P_{Cl} cm/sec	$[Cl]_i$ mM	α^i_{Cl} mM	E_{Cl} mM	Reference
Sheep Purkinje	-74.1	2.4×10^{-8}		13.8	-51.7	Vaughan-Jones, 1977
Sheep Purkinje	—	—	36	—	—	Carmeliet & Bosteels, 1969
Cow Purkinje	-75	3.4×10^{-8}#	—	—	—	Carmeliet & Verdonck, 1977
Frog Ventricular	-90.4	—	—	17.6	-40.2	Ladle & Walker, 1975
Frog Atrial	-89.8	—	—	17.3	-36.3	Ladle & Walker, 1975
Frog Sinus Venosus	-55.7	—	—	35.0	-25.0	Ladle & Walker, 1975
Cat Papillary	—	—	34 21.3##	—	—	Carmeliet & Janse, 1965
Rat Atrial	—	—	19.5##	—	—	Lamb, 1961
Rabbit Atrial	—	—	37	—	—	DeMello, 1963

α^i_{Cl}: Internal Cl^- activity. Unless otherwise specified, the $[Cl]_i$ is based on total tissue Cl^- analyses. The α^i_{Cl} was measured with chloride-selective microelectrodes.

#The P_{Cl}/P_K ratio was 0.205. A P_{Cl}/P_K ratio of 0.11 was obtained by Fozzard and Lee (1976) for rabbit ventricular muscle, and the ratio of g_{Cl} / g_K was 0.17.

##$[Cl]_i$ measured from radioisotope washout.

Appendix V

DERIVATION OF NERNST EQUATION

The Nernst equation may be derived from the general equation for the free energy change (ΔG^c) resulting from both **osmotic work** and **electrical work** for transporting 1 mol of cation (c^+) across a membrane

$$\Delta G^c = RT \ln \frac{[c^+]_i}{[c^+]_o} + zF\,E_m \qquad \text{eq. A-1}$$

where R is the gas constant; T is absolute temperature; $[c^+]_i$ and $[c^+]_o$ are the internal and external c^+ concentrations, respectively; z is the valence; F is the **Faraday constant**; and E_m is membrane potential. The first term on the right side of this equation, $RT\ \ln([c^+]_i/[c^+]_o)$, gives the osmotic work for transporting a mole of particles across the membrane against a **concentration gradient**. The second term, zFE_m, gives the electrical work for transporting 1 mol of charged particles across the membrane against an electrical gradient. The sum of these two terms then gives the total work required. At equilibrium, the change in free energy for moving one or only a few particles across the membrane must be zero ($\Delta G = 0$). Therefore

$$0 = RT \ln \frac{[c^+]_i}{[c^+]_o} + zF\,E_m \qquad \text{eq. A-2}$$

and

$$zF\,E_m = -RT \ln \frac{[c^+]_i}{[c^+]_o} \qquad \text{eq. A-2a}$$

or

$$E_m = (-RT/zF)\ \ln \frac{[c^+]_i}{[c^+]_o} \qquad \text{eq. A-2b}$$

which is the **Nernst equation**.

Since the Faraday constant (F) is equal to the charge on an electron (e, in coul) times **Avogadro's number** (N_A, number of ions per mol), the following is true

$$\frac{RT}{F} = \frac{RT}{N_A q_e} = \frac{kT}{q_e} \qquad \text{eq. A-3}$$

where k (the **Boltzmann constant**) is equal to the gas constant divided by Avogadro's number (N_A), i.e., the energy (in joul) of an ion per degree Kelvin, and q_e is the charge (in coul) on an electron.

Chapter VI

DETERMINATION OF RESTING POTENTIAL

A. Modified Nernst Equation

The basic Nernst equation has been modified in several ways for special situations. For example, in the concentration cell illustrated in Chapter V (Fig. V-1), a cell in which a single salt (both ions of same valence) is distributed across the membrane at two different concentrations, the PD developed across the membrane (E_m) can be calculated from the following equation (of course, this equation does not apply to biological, or living, cells in which several salts are distributed across the cell membrane)

$$E_m = \frac{U_c - U_a}{U_c + U_a} \frac{-61\,\text{mV}}{z} \log \frac{[\text{salt}]_1}{[\text{salt}]_2} \qquad \text{eq. 1}$$

where $[\text{salt}]_1$, and $[\text{salt}]_2$ are the concentrations of the salt on side 1 and side 2, and U_c and U_a are the mobilities of the cations and anions, respectively, through the membrane. Thus, when the **mobility** (or permeability) of the cation and anion are equal ($U_c = U_a$), E_m is zero, regardless of the equilibrium potentials. When the anion is impermeable ($U_a = 0$), then the mobility fraction in Equation 1 becomes 1.0, and the equation reduces to the simple Nernst equation. When the cation is impermeable ($U_c = 0$), the fraction becomes -1, and the same numerical value of E_m is produced, but of the opposite sign. Equation 1 can be used to calculate E_m for any combination of U_a and U_c. For example, if $U_a = 0.5\ U_c$, then for the problem illustrated in Chapter V (Fig. V-1), E_m is about -20 mV (side 1 negative). Thus, Equation 1 relates the membrane potential to the relative mobilities.

B. Constant Field Equation

An important modification of the Nernst equation in common use for calculating the membrane potential, or more importantly for determining the P_{Na} / P_K ratio, is the Goldman-Hodgkin-Katz (Goldman, 1943; Hodgkin & Katz, 1949) **constant field equation**

$$E_m = \frac{-RT}{F} \ln \frac{P_K[K^+]_i + P_{Na}[Na^+]_i + P_{Cl}[Cl^-]_o}{P_K[K^+]_o + P_{Na}[Na^+]_o + P_{Cl}[Cl^-]_i} \qquad \text{eq. 2}$$

where P_K, P_{Na}, and P_{Cl} are the membrane permeabilities for K^+, Na^+, and Cl^-, respectively. The P_K $[K]_i$ product, for example, is given in the units of a flux (mol/sec per cm^2), but the entire right-hand term (fraction) is dimensionless. Plugging in the numerical values for the constants and converting from natural logarithm, or ln, to logarithm to the base 10, or log ($\ln N = 2.3 \log N$), and dividing the right-hand term by P_K gives

$$E_m = -61\,mV \log \frac{[K^+]_i + \frac{P_{Na}}{P_K}[Na^+]_i + \frac{P_{Cl}}{P_K}[Cl^-]_o}{[K^+]_o + \frac{P_{Na}}{P_K}[Na^+]_o + \frac{P_{Cl}}{P_K}[Cl^-]_i} \qquad \text{eq. 2a}$$

E_m expressed in mV. Again, the right-hand term (fraction) is dimensionless.

Any ion that is passively distributed, i.e., not actively pumped, however, cannot determine the resting potential, because the ion's distribution must follow the resting potential. Therefore, when Cl^- is passively distributed, it drops out of consideration. (This is not to say that transient net movements of Cl^- inward or outward do not influence E_m, because they certainly do; it is only that the steady-state resting potential cannot be determined by Cl^-.) Therefore, Equation 2a can be reduced to

$$E_m = -61\text{ mV}\ \log \frac{[K^+]_i + \frac{P_{Na}}{P_K}[Na^+]_i}{[K^+]_o + \frac{P_{Na}}{P_K}[Na^+]_o} \qquad \text{eq. 2b}$$

Equation 2b is one of the most useful forms of the constant field equation, because if E_m is measured, and if the internal and external ion concentrations are known, the P_{Na}/P_K ratio can be calculated. Thus for given ionic gradients, the resting potential is determined by the P_{Na}/P_K ratio (i.e., the relative permeabilities of the cell membrane to Na^+ and K^+) and not by the absolute permeabilities. For simplicity, Ca^{2+} is ignored here as a factor contributing to the resting potential. The relationship between the permeability coefficient for an ion (P_i) and the membrane conductance for that ion (g_i) is given in the Appendix A to this chapter, as are the constant field equations for the ionic currents.

Figure VI-1 gives the expected resting potential as a function of the P_{Na}/P_K ratio for a muscle cell (having a $[K]_i$ of 150 mM) bathed in normal Ringer's solution ($[K]_o$ of 4 mM, $[Na]_o$ of 150 mM), assuming two different $[Na]_i$ values (15 mM and 4 mM). As can be seen in the figure, a P_{Na}/P_K ratio of 0.1 gives a resting potential close to -60 mV, whereas a ratio of 0.01 gives close to -85 mV. Some muscle cells, e.g., smooth muscle and young embryonic myocardial cells, have a low resting potential (at a $[K]_o$ of 4 mM) of about -50 mV, presumably because of a high P_{Na}/P_K ratio of about 0.15 rather than because of altered ionic gradients. The P_{Na}/P_K ratio can

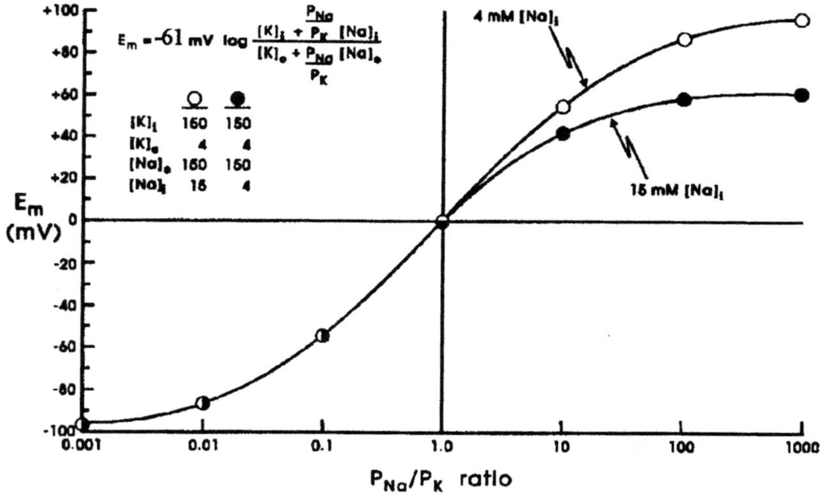

FIGURE VI-1. Theoretical curves calculated from Goldman constant-field equation (inset) of membrane potential (E_m) for different ratios of P_{Na}/P_K. Calculations made for a $[K]_i$ of 150 mM, $[K]_o$ of 4 mM $[Na]_o$ of 150 mM, and a $[Na]_i$ either of 15 mM (solid circles) or of 4 mM, (open circles). Resting potential is about -85 mV for a P_{Na}/P_K ratio of 0.01 and about -65 mV for a ratio of 0.1. Action potential overshoot predieted, ignoring any Cl^- short-circuiting, for the various P_{Na}/P_K ratios is also given.

be high because of either a high P_{Na} or a low P_K, or both; in most cases the main reason appears to be a low P_K (see chapter X on *Developmental Changes in* K^+ *Permeability*).

One advantage Equation 2b has over the chord conductance equation (discussed in the subsequent subsection C) is that it nicely accounts for the bend at low $[K]_o$ in the E_m vs. log $[K]_o$ curves (see subsection D *Resting Potential as Function of* $[K]_o$ and Fig. VI-2). Figure VI-2 presents theoretical curves, calculated from Equation 2b, for E_m vs. log $[K]_o$ at different P_{Na} / P_K ratios. The higher the P_{Na} / P_K ratio, the greater the deviation is from a straight line as $[K]_o$ is lowered. From Equation 2b one can deduce that as $[K]_o$ is lowered and $[Na^+]_o$ is concomitantly elevated, the denominator of the right-hand term becomes more and more dominated by the Na^+ term, for any given P_{Na}/P_K ratio. Because the numerator is relatively fixed, E_m is more and more influenced by E_{Na} as $[K]_o$ is lowered more and more; thus this relationship accounts for the deviation from the straight line for E_K.

Although the $P_{Na}/P_K = 0.05$ curve is almost linear at high $[K]_o$ with a slope of 60 mV/decade, the membrane does not necessarily become purely K^+ selective, as is often stated, because for these theoretical calculations the P_{Na}/P_K ratio was held constant over the entire $[K]_o$ range. There is some evidence, however, that P_K itself increases as $[K]_o$ increases, but this effect was not taken into consideration in Figure VI-2. (It is well recognized that the g_K is a function of $[K]_o$, namely, that $g_K \propto P_K [K]_o$, as discussed in the Appendix VI-A to this chapter.) Finally, the increased bending for the higher P_{Na}/P_K ratios can again be seen from Equation 2b; at a given $[K]_o$ (e.g., 4 mM) the denominator is more and more dominated by the Na^+ term as the P_{Na}/P_K ratio is increased more and more.

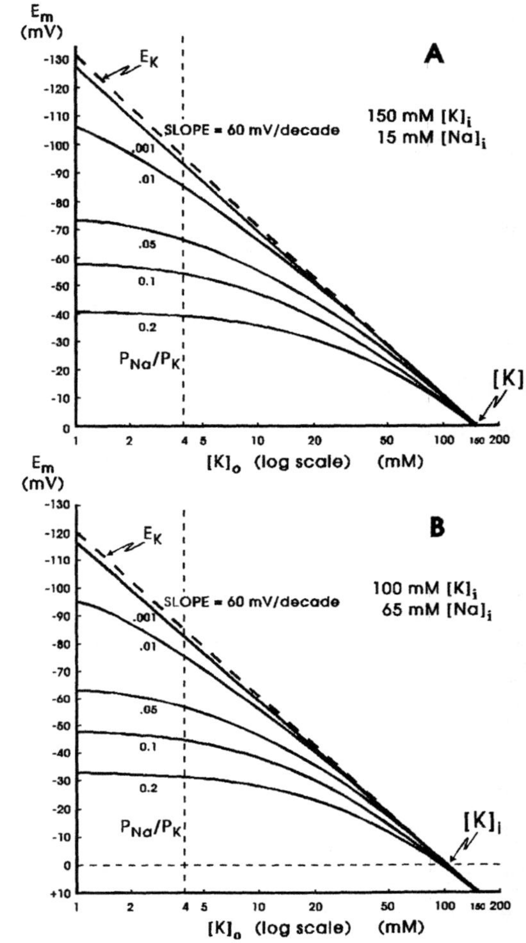

FIGURE VI-2. Theoretical curves calculated from Goldman constant field equation for resting potential (E_m) as a function of $[K]_o$. Family of curves is given for various P_{Na}/P_K ratios (0.001, 0.01, 0.05, 0.1, and 0.2). K^+ equilibrium potential (E_K) calculated from the Nernst equation (broken straight line). **A:** curves calculated for a $[K]_i$ of 150 mM and a $[Na]_i$ of 15 mM. **B:** Curves calculated for a $[K]_i$ of 100 mM and a $[Na]_i$ of 65 mM. Sum of $[K]_i$ and $[Na]_i$ held constant. Calculations made holding $[K]_o + [Na]_o$ constant at 154 mM; i.e., as $[K]_o$ was elevated, $[Na]_o$ was lowered by an equimolar amount. Change in P_K as a function of $[K]_i$, was not taken into account for these calculations. Point at which E_m is zero gives $[K]_i$. Note that the potential reverses in sign when $[K]_o$, exceeds $[K]_i$, in B.

The order of selectivity of the resting membrane for the alkali metal ions is generally in the following sequence, from the highest permeability to the lowest: $K^+ > Rb^+ > Cs^+ > Na^+ > Li^+$. For example, the relative permeabilities (assigning $P_K = 1$) in squid giant axon (Baker et al., 1969b) are

P_K >	P_{Rb} >	P_{Cs} >	P_{Na} >	P_{Li}
1.0	0.69	0.19	0.17	0.12

In frog sartorius (Mullins, 1961) the values are

P_K >	P_{Rb} >	P_{Cs} >	P_{Na}
1.0	0.54	0.11	0.04

The P_{Ba}/P_{Ca} ratio in frog sartorius is about 3 (Mullins, 1961).

So far in our discussion, Ca^{2+} has been ignored. It can be demonstrated, however, that Ca^{2+} has only a negligible effect on the resting potential, even if it has a permeability equal to, or 10 times greater than Na^+. This is because of the relatively low extracellular and intracellular concentrations of free Ca^{2+} ion, compared to those of the K^+ or Na^+ ions. Following is a modified version of the Goldman constant-field equation which includes a Ca^{2+} term*

$$E_m = -60 \text{ mV} \log \frac{(B-A)+\sqrt{y}}{2(A-4P_{Ca}[Ca]_i)} \qquad \text{eq. 3}$$

where

* This equation was kindly provided by Prof. D.E. Goldman.

$$A = P_K[K]_i + P_{Na}[Na]_i$$
$$B = P_K[K]_o + P_{Na}[Na]_o$$
$$y = (B - A)^2 + 4(A + 4P_{Ca}[Ca]_i)(B + 4P_{Ca}[Ca]_o)$$

For simplicity, the analogous Cl^- terms ($+P_{Cl}[Ca]_o$ in the definition of A and $+P_{Cl}[Cl]_i$ term in B) have been omitted, assuming Cl^- to be passively distributed.

Calculations made from Equation 3 are given in Tables VI-1 and VI-2. These calculations demonstrate some interesting points: (a) For the same permeabilities ($P_{Ca} = P_{Na}$), Ca^{2+} has a much less effect on E_m than does Na^+, because of the lower Ca^{2+} concentrations and because of the square root function for the Ca^{2+} concentrations. For example, the depolarization produced by taking into account the Ca^{2+} ion is only +0.4 mV for a P_{Na}/P_K ratio of 0.1 (Table VI-1). (b) Even when P_{Ca} is set equal to 10 times P_{Na}, the effect of Ca^{2+} on the resting potential is still relatively small (e.g., +3.5 mV for a P_{Na}/P_K ratio of 0.01, and +8.0 mV for a P_{Na}/P_K ratio of 0.1). (c) the effect of Ca^{2+} is somewhat greater when the P_{Na}/P_K ratio is higher (as in cardiac nodal cells or smooth muscle). (d) The effect of taking into account Ca^{2+} is considerably greater at low $[K]_o$ values than at high $[K]_o$ values (Table VI-2).

Thus, these calculations support the view that Ca^{2+} can be virtually ignored in discussion of the ionic basis of the resting potential. This agrees with the well-known fact that a variation in [Ca] throughout a relatively wide range has a negligible effect on the resting potential (see, for example, Sperelakis, 1972a). Further, these conclusions have implications about the relative importance of Na^+ vs. Ca^{2+} background currents (inward) during genesis of the pacemaker potential (concomitant with the decrease in g_K and I_K). Finally, it should be emphasized that, for example, when P_{Ca} =

TABLE VI-1

Calculation of resting potential for different values of Ca^{2+} permeability using modified version of the Goldman constant field equation.

P_K	P_{Na}	P_{Ca}	$\frac{P_{Na}}{P_K}$	$\frac{P_{Ca}}{P_K}$	Resting E_m mV
10^{-6}	10^{-8}	0	0.01	0.0	-86.2
10^{-6}	10^{-8}	10^{-9}	0.01	0.001	-86.1
10^{-6}	10^{-8}	10^{-8}	0.01	0.01	-85.8
10^{-6}	10^{-8}	10^{-7}	0.01	0.1	-82.7
10^{-6}	10^{-8}	10^{-6}	0.01	1.0	-64.0
10^{-6}	10^{-8}	10^{-5}	0.01	10.0	-23.0
10^{-7}	10^{-8}	0	0.1	0.0	-54.1
10^{-7}	10^{-8}	10^{-9}	0.1	0.01	-54.0
10^{-7}	10^{-8}	10^{-8}	0.1	0.1	-53.1
10^{-7}	10^{-8}	10^{-7}	0.1	1.0	-46.1
10^{-7}	10^{-8}	10^{-6}	0.1	10.0	-19.0
10^{-6}	10^{-7}	10^{-7}	0.1	0.1	-53.1
10^{-5}	10^{-7}	10^{-7}	0.01	0.01	-85.8

Constant conditions: $[K]_o$ = 4 mM; $[K]_i$ = 150 mM; $[Na]_o$ = 150 mM; $[Na]_i$ = 15 mM; $[Ca]_o$ = 2 mM; $[Ca]_i$ = 10^{-4} mM; and E_K = -94 mV. Lowering $[Na]_i$ 10-fold to 1.5 mM had only a slight effect, reducing the calculated E_m by less than 0.2 mV when P_{Ca} was made equal to P_{Na}.

TABLE VI-2

Calculation of resting membrane potential as a function of external K^+ concentration for different values of Ca^{2+} permeability using a modified version of the Goldman constant-field equation.

$[K]_o$ M	$P_K = 10^{-6}$ $P_{Na} = 10^{-8}$				$P_K = 10^{-7}$ $P_{Na} = 10^{-8}$			
	P_{Ca}				P_{Ca}			
	0	10^{-9}	10^{-8}	10^{-7}	0	10^{-9}	10^{-8}	10^{-7}
1	-106.4	-106.3	-105.6	-99.4	-58.1	-58.0	-57.0	-48.8
2	-97.8	-97.7	-97.2	-92.6	-56.7	-56.6	-55.6	-47.9
4	-86.2	-86.1	-85.8	-82.7	-54.1	-54.0	-53.1	-46.1
8	-72.0	-72.0	-71.8	-70.0	-49.6	-49.5	-48.8	-42.8
10	-67.1	-67.1	-66.9	-65.4	-47.6	-47.5	-46.9	-41.3
20	-50.8	-50.8	-50.7	-50.0	-39.4	-39.3	-38.9	-34.9
40	-33.7	-33.7	-33.7	-33.3	-28.2	-28.1	-27.9	-25.4
80	-16.2	-16.2	-16.1	-16.0	-14.3	-14.3	-14.2	-12.9
150	-0.02	-0.02	-0.01	-0.05	-0.2	-0.2	-0.1	+0.5

Constant conditions: $[K]_i$ = 150 mM; $[Na]_i$ = 15 mM; $[Ca]_o$ = 2 mM; $[Ca]_i$ = 10^{-4} mM; $[K]_o$ + $[Na]_o$ held constant at 154 mM. Lowering $[Na]_i$ 10-fold to 1.5 mM had only a slight effect.

P_K , g_{Ca} does not equal g_K , because of the concentration differences. To calculate g_K from a given P_K , one must use the appropriate equation (given in the Appendix VI-A to this chapter) that takes into account the concentrations and membrane potential. If $g_{Ca} = g_K$ (ignoring Na^+), then the membrane potential is halfway between E_K and E_{Ca} , as indicated by the chord conductance equation; that is, in this example, Ca^{2+} has a much greater effect on the resting potential.

C. Chord Conductance Equation

An alternative method of approximating the membrane resting potential (E_m) is by use of the **chord conductance equation**. The word *chord* refers to a straight line connecting two points on a curve. Specifically, here chord refers to the **average slope** of a nonlinear steady-state voltage-current curve, i.e., a straight line from any point on the curve through the origin (zero applied current). It contrasts with **slope conductance**, which is the tangent at any point on the curve. The equation is written as

$$E_m = \frac{g_K}{\Sigma g} E_K + \frac{g_{Na}}{\Sigma g} E_{Na} + \frac{g_{Cl}}{\Sigma g} E_{Cl} + \frac{g_{Ca}}{\Sigma g} E_{Ca} + \cdots \qquad \text{eq. 4}$$

where g_K, g_{Na}, g_{Cl}, and g_{Ca} are the membrane conductances for K^+, Na^+, Cl^-, and Ca^{2+}, respectively, and Σg is the total conductance (sum of all ionic partial conductances). Thus, for example, the ratio of g_K/Σ_g is the **relative or fractional conductance** for K^+.

The chord conductance equation can conveniently take into account all ions, including divalent cations, that are distributed unequally across the membrane and that have a finite conductance. The ions important to membrane potentials (including action potentials, postsynaptic potentials, and receptor potentials) are K^+, Na^+, Cl^-, and Ca^{2+}; but as discussed in the previous section, Cl^- cannot help to determine the resting potential if it is passively distributed. Thus, Equation 4 can be rewritten, omitting the Cl^- term

$$E_m = \frac{g_K}{\Sigma g} E_K + \frac{g_{Na}}{\Sigma g} E_{Na} + \frac{g_{Ca}}{\Sigma g} E_{Ca} \qquad \text{eq. 4a}$$

For simplicity, we will ignore the Ca^{2+} term also, giving

$$E_m = \frac{g_K}{g_K + g_{Na}} E_K + \frac{g_{Na}}{g_K + g_{Na}} E_{Na} \qquad \text{eq. 4b}$$

where Σg is now equal to $g_K + g_{Na}$.

Another advantage of the chord conductance equation is that it can be derived simply from Ohm's law and from circuit analysis for a condition when net current is zero, and a derivation of Equation 4b is given in the Appendix VI-B to this chapter. The equation holds true whenever the net current across the membrane is zero, as for the resting potential. The chord conductance equation is useful for giving the membrane potential when the ion conductances and distributions are known or for giving the fractional conductances when the ion distributions are known and E_m is measured. For example, if g_K is $15 \times g_{Na}$, and E_K = -94 mV and E_{Na} = +60 mV, then the resting potential is

$$\begin{aligned} E_m &= 15/16(-94\ \text{mV}) + 1/16(+60\ \text{mV}) \qquad \text{eq. 4c} \\ &= -88.1 + 3.75 \\ &= -84.3\ \text{mV} \end{aligned}$$

At the neuromuscular junction, the neurotransmitter acetylcholine opens up the gates of many ionic channels that allow both Na^+ and K^+ to pass through equally well (that is, $g_{Na} = g_K$). Hence, the potential that the postsynaptic membrane tends to seek when maximally activated (i.e., the equilibrium potential or so-called **reversal potential** for the end-plate potential, EPP) is

$$E_{EPP} = 1/2(-94 \text{ mV}) + 1/2(+60 \text{ mV}) \quad \text{eq. 4d}$$
$$= -17 \text{ mV}$$

The peak of the end-plate potential (EPP) does not quite reach -17 mV, depending on the equivalent circuit for the end plate (see Appendix VI-C to this chapter for the circuit analysis). If at the peak of the action potential in skeletal muscle g_{Na} is 9 × g_K, then the peak overshoot potential ($+E_{max}$) is close to

$$+E_{max} = 1/10(-94 \text{ mV}) + 9/10(+60 \text{ mV}) \quad \text{eq. 4e}$$
$$= -9.4 + 54.0$$
$$= +44.6 \text{ mV}$$

Hence, the chord conductance equation gives a close approximation of the membrane potential expected for various relative conductances. The equation indicates that the membrane potential is dominated by the diffusion potential for that ion whose conductance is greatest. Circuit analysis explains why the battery having the resistance is the battery that dominates the membrane potential (see Appendix VI-C to this chapter).

One disadvantage of the chord conductance equation is that it gives nearly a straight line for the E_m vs. log $[K]_o$ plot (actually a slight bend in the opposite direction at low $[K]_o$). In contrast, the constant field equation gives the complete bending of the curves (for different P_{Na}/P_K ratios) (see section B *Constant Field Equation* in this chapter). In addition, because when $[K]_o = 0$, E_K approaches infinity, the chord conductance equation does not hold true at very low $[K]_o$ values; in contrast, the constant field equation does apply (at least in principle). The g ratios in the chord conductance equation are equivalent to transport numbers, but the ratios vary with concentration. The complex relationship between conductance and permeability is also given in the Appendix VI-A to this chapter. Conductance is dependent on permeability, on the ion concentrations on both sides of the membrane, and on the membrane potential.

D. Resting Potential as Function of $[K]_o$

The resting potential (E_m) decreases as the external K^+ concentration ($[K]_o$) is elevated, according to the relationship indicated in Figure VI-2. In Figure VI-2, the theoretical calculations of E_m from the Goldman constant field equation are plotted against log $[K]_o$ for cells with different ratios of P_{Na}/P_K. (These curves were calculated assuming that the P_{Na}/P_K ratio stayed constant at all $[K]_o$ levels, a condition not always true.) In this plot the elevation in $[K]_o$ is obtained by equimolar substitution for Na^+, in order to keep the tonicity constant; hence $[K]_o + [Na]_o$ is held constant at 154 mM. Let us focus first on the curve for a P_{Na} / P_K ratio of 0.01 and on the nearly superimposed data points found for adult myocardial cells. Elevation of $[K]_o$ obviously causes the K^+ concentration gradient to be diminished. In acute or even chronic (Pappano & Sperelakis, 1969) experiments the $[K]_i$ does not increase substantially in the presence of an elevated $[K]_o$. Therefore from the Nernst equation, E_K is decreased.

The straight line in Figure VI-2 shows how E_K varies for a myocardial cell with a $[K]_i$ of 150 mM as $[K]_o$ is elevated or lowered. ($[K]_i$ is often estimated by extrapolating the E_m vs. log $[K]_o$ curve to zero potential. (This extrapolation assumes that the activity coefficients for K^+ outside, $\alpha_K{}^o$ and inside, $\alpha_K{}^i$, are equal and that the membrane has no asymmetry or residual potential when both sides are bathed by identical solutions.) This line has a slope of 60 mV/10-fold (decade) change in $[K]_o$, as predicted from the Nernst equation. As can be seen in Figure VI-2, the slope of the curve for the myocardial cell with a P_{Na}/P_K ratio of 0.01 in $[K]_o$ solutions greater than 10 mM is also nearly 60 mV/decade; i.e., the slope parallels and reflects the change in E_K. Below a $[K]_o$ of 10 mM, the curve bends, deviating from E_K. This amount of deviation is nearly exactly predicted from the Goldman constant field equation, as mentioned in section B *Constant Field Equation* in this chapter, because as $[K]_o$ is lowered and $[Na]_o$ is concomitantly elevated, the composite term in the denominator of the logarithmic expression is more and more influenced by the Na^+ term, and correspondingly less influenced by the K^+ term, for any given P_{Na}/P_K ratio. Thus, at a normal $[K]_o$ of 4 mM, the resting E_m (about -80 mV) is about 14 mV less than E_K (about -94 mV).

Besides the change in E_K as $[K]_o$ is varied, the K^+ conductance of the membrane (g_K) is also changed. The g_K is a function of $[K]_o$, because elevation of $[K]_o$, increases g_K (see the Appendix VI-A to this chapter). Thus, membrane conductance for an ion is dependent on the permeability of the membrane to that ion as well as on the concentrations of that ion on both sides of the membrane, i.e., on the number of ions available to carry current. In addition, there is evidence that P_K itself is also affected by $[K]_o$, because P_K is lower in lower $[K]_o$ (e.g., see Carmeliet et al., 1976). Therefore, a second factor that contributes to the E_m vs. log $[K]_o$ curve is the change in g_K and P_K as $[K]_o$ is varied. This increase in g_K and P_K acts, however, to hyperpolarize the membrane toward the calculated E_K at any given $[K]_o$ level. (For example, an increase in P_K by acetylcholine hyperpolarizes atrial cells and sinoatrial nodal cells toward E_K at any $[K]_o$ level.) In other words, the amount of depolarization produced by any given increment in $[K]_o$ is slightly less, because of this second factor that increases g_K and P_K. This factor may be partly responsible for the hyperpolarization observed (Sperelakis & Lehmkuhl, 1966; DeHaan & Gottlieb, 1968) in some heart cells when $[K]_o$ is elevated to 10-15 mM (stimulation of an electrogenic Na^+ pump at an optimal $[K]_o$ could also be involved), and this effect could be more prominent at the normal $[K]_o$ values cells that have a low P_K (therefore high P_{Na}/P_K ratio and low resting potential).

Nodal cells of the heart, smooth muscle cells of blood vessels and viscera, and young embryonic ventricular cells (see Chapter X on *Developmental Changes in K^+ Permeability*) have high P_{Na}/P_K ratios of 0.1-0.2, mainly because of a relatively low P_K (and not necessarily a high P_{Na}). These cells appear to have a normally high $[K]_i$ level, and therefore their observed low resting potentials of -40 to -60 mV in normal K^+-containing solutions (4 mM) cannot be explained on the basis of a low E_K. As shown in the theoretical curves in Figures VI-1 and VI-2, such cells should have a low resting potential because of the high P_{Na}/P_K ratio; i.e., the membranes of these cells are less selective for K^+ than are those of atrial and ventricular myocardial cells or skeletal muscle twitch fibers. The curves have a much greater deviation from E_K and are much flatter in low $[K]_o$. Further, as can be seen in Figure VI-2, a given increment in $[K]_o$ does not produce as much depolarization. The curves are not linear in elevated $[K]_o$, but they continuously bend. The average slope, between 15 and 150 mM $[K]_o$ is about 30 mV/decade for cells with a P_{Na}/P_K ratio of 0.2, compared to 60 mV/decade for cells with a P_{Na}/P_K ratio of 0.01.

Sometimes as $[K]_o$ is elevated (e.g., to 10 mM), a slight to moderate hyperpolarization is produced; elevation to more than about 15 mM, however, produces the usual depolarization (Fig. VI-3). Such a phenomenon is observed more frequently in cells with a high P_{Na}/P_K ratio (due to a low P_K) and therefore with a low resting potential, such as cultured heart cells (Sperelakis & Lehmkuhl, 1966) and dystrophic skeletal muscle fibers (Sellin & Sperelakis, 1978). This hyperpolarization could be explained by stimulation of an electrogenic Na^+ pump when $[K]_o$ is elevated (see Chapter IX on *Electrogenic Sodium Pump Potentials*). On the other hand, the hyperpolarization could also be explained by an increase in P_K and g_K produced by the increase in $[K]_o$ (Sperelakis & Lehmkuhl, 1966).

At a fixed $[K]_o$, a variation in $[Na]_o$ by substitution with an impermeant molecule such as sucrose (for retention of isosmolarity) also influences the resting potential, as expected. From the constant field equation, one can see that for any given P_{Na}/P_K ratio, as the $[Na]_o$ is lowered while $[K]_o$ is held constant, the denominator of the logarithmic term becomes more dominated by the K^+ term. Hence, hyperpolarization should occur toward E_K. Another way to consider this is that lowering $[Na]_o$ should lower Na^+ influx, which is a depolarizing current; hence hyperpolarization should occur. Such hyperpolarization is prominent in cultured heart cells that have a low P_K , and hence a low resting potential (Pappano & Sperelakis, 1969).

Variation in $[Cl^-]_o$ does not affect the steady-state resting potential in cells in which Cl^- is passively distributed in accordance with the resting potential, i.e., not actively pumped. Under such conditions, there is no net movement of Cl^-. On the other hand, when net movement of Cl^- ions occurs, this

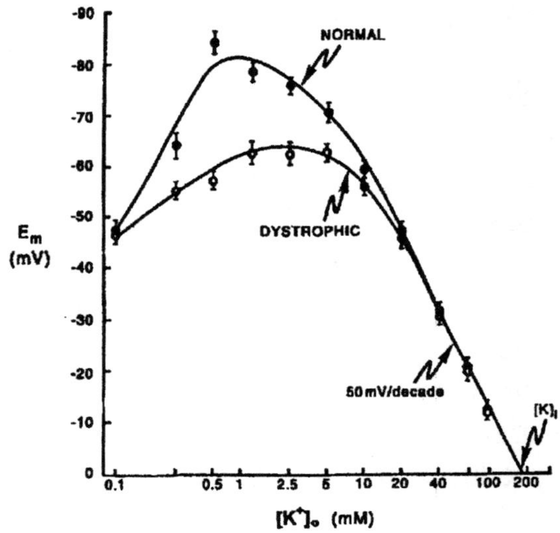

FIGURE VI-3. The resting membrane potential (E_m) of mouse skeletal muscle (extensor digitorum longus muscle) is plotted as a function of the extracellular K^+ concentration ($[K]_o$) on a logarithmic scale. **Solid circles:** fibers from normal mice. **Open circles:** fibers (innervated) from dystrophic mice. Data points plotted mean ± SE. The lines drawn through the data for 20 mM [K]o and above were fitted by the least squares method, and they have a slope of 50 mV/decade for both the dystrophic muscles and the normal muscles. Extrapolation of the lines to zero potential gives the internal K^+ concentration ($[K]_i$). (Modified from Sellin and Sperelakis, 1978.)

movement does transiently affect the membrane potential. For example, when skeletal muscle is placed in a Cl^--free solution, a net Cl^- efflux produces depolarization with spontaneous action potentials and twitches; but when the net Cl^- efflux ceases, the resting potential returns to the original -90 mV. Reintroduction of Cl^- into the bath to produce a net Cl^- influx causes a transient, pronounced hyperpolarization. The magnitude and time course of these potential changes depend, of course, on P_{Cl}; the larger the P_{Cl}, the greater the potential change is. During the skeletal fiber action potential, E_m is different from E_{Cl}, and there is a net driving force for Cl^- influx that helps to speed up the repolarizing phase terminating the action potential.

In skeletal muscles that have a high P_{Cl}, often times $[Cl]_o$ is reduced also (in addition to Na^+) as $[K]_o$ is elevated (by substitution with an impermeant anion such as methyl sulfonate) in such a fashion that the product of $[K]_o$, and $[Cl]_o$ is held constant ($[K]_o \times [Cl]_o$ = constant, e.g., at 480 mM^2). This $[Cl]_o$ reduction is done in order to prevent a net Cl^- influx that would otherwise occur. As the resting potential is diminished in high $[K]_o$, the potential that was effectively excluding Cl^- from the cell decreases (Cl^- passively distributed), and hence the cell gains Cl^-. Na^+ tends to accompany the Cl^- for electroneutrality of the bulk myoplasm, but some of the Na^+ is, in turn, pumped back out and K^+ is pumped in; therefore, effectively the cell should gain K^+, Na^+, Cl^-, and H_2O when swollen by this means. The internal ion concentrations, however, are not greatly increased, because of the dilution. A net Cl^- influx exerts a transient hyperpolarizing action and increases the total number of osmotic particles in the cell, so that a net water influx and concomitant swelling of the cells occurs.

Therefore, if $[Cl]_o$ is lowered as $[K]_o$ is elevated, the diffusion force for Cl^- is reduced in proportion to the reduced electrical force, thus keeping the net electrochemical driving force constant at zero. For this purpose, E_{Cl} must be always made equal to E_m, so that a net driving force for Cl^- never exists. This condition holds true when the product of $[K]_o$ and $[Cl]_o$ are held constant, as shown in the next derivation. Using the approximation that the resting potential is equal to E_K, then:

$$E_m = E_K = E_{Cl} \qquad \text{eq. 5}$$

Substituting for E_K and E_{Cl} from the Nernst equation gives:

$$-61\ \text{mV} \log \frac{[K^+]_i}{[K^+]_o} = -61\ \text{mV} \log \frac{[Cl^-]_o}{[Cl^-]_i} \qquad \text{eq. 5a}$$

or

$$\frac{[K^+]_i}{[K^+]_o} = \frac{[Cl^-]_o}{[Cl^-]_i} \qquad \text{eq. 6}$$

Hence,

$$[K^+]_o \times [Cl^-]_o = [K^+]_i \times [Cl^-]_i \qquad \text{eq. 6a}$$

Because $[K]_i$ is relatively fixed and because we wish to hold $[Cl^-]_i$ constant so that there is no net Cl^- influx, then

$$[K^+]_i \times [Cl^-]_i = \text{constant} \qquad \text{eq. 6b}$$

Therefore

$$[K^+]_o \times [Cl^-]_o = \text{constant} \qquad \text{eq. 6c}$$

Equation 6c then gives the justification for lowering [Cl] in the bathing solution as [K] is elevated.

Appendix VI-A

CONSTANT FIELD EQUATION

As stated in Chapter IV on *Diffusion and Permeability*, for uncharged molecules or for ions in the absence of an electrical field, the influx (J_i) is a product of the permeability coefficient (P) times the external concentration (C_o), and the efflux (J_o) is a product of P times the internal concentration (C_i)

$$J_i = -P\ C_o \qquad \text{eq. AA-1}$$

$$J_o = P\ C_i \qquad \text{eq. AA-2}$$

For charged particles in the presence of an electrical field, the concentration term must be multiplied by a factor that takes into account the effect of the electrical field on the passage of the ion through the membrane. For univalent ions, this factor (see Katz, 1966) is

$$\text{factor} = \frac{VF\ /\ RT}{1 - e^{-VF/RT}} \qquad \text{eq. AA-3}$$

where V is the PD across the membrane, the sign for V being *positive* when the ion movement is facilitated by the field and *negative* when the ion movement is opposed. Thus, for inward movement of Na^+ or K^+ across the resting membrane (or for outward movement of Cl^-), this **assistance factor** is

$$\text{assistance factor} = \frac{+80\ \text{mV}/26\ \text{mV}}{1 - e^{-(+80\ \text{mV}/26\ \text{mV})}} \qquad \text{eq. AA-3a}$$

$$= \frac{3.077}{1 - \left(1/e^{3.077}\right)}$$

$$= \frac{3.077}{1 - (1/21.693)} = \frac{3.077}{0.9539}$$

$$= 3.226$$

For outward movement of K^+ or Na^+ (or for inward movement of Cl^-), the **hindrance factor** is

$$\text{hindrance factor} = \frac{-80\ \text{mV}/26\ \text{mV}}{1 - e^{-(-80\ \text{mV}/26\ \text{mV})}} = \frac{-3.077}{1 - \left(e^{3.077}\right)} \qquad \text{eq. AA-3b}$$

$$= \frac{-3.077}{1 - 21.693} = \frac{-3.077}{-20.693}$$

$$= 0.1487$$

Note that these factors are dimensionless. Hence, the equations for passive K^+ influx and efflux at the resting potential are

$$J_i^{\text{K}} = -P_{\text{K}}[\text{K}^+]_o \frac{VF}{RT} \frac{1}{1 - e^{VF/RT}} \qquad \text{eq. AA-4}$$

$$J_o^{\text{K}} = P_{\text{K}}[\text{K}^+]_i \frac{VF}{RT} \frac{1}{1 - e^{-VF/RT}} \qquad \text{eq. AA-5}$$

Thus, K^+ influx increases with an increase in P_K and $[K]_o$ and with hyperpolarization. If P_K is assumed to be 1×10^{-6} cm/sec, then

$$J_i^{\mathrm{K}} = \left(1 \times 10^{-6} \frac{\mathrm{cm}}{\mathrm{sec}}\right) \left(4 \times 10^{-6} \frac{\mathrm{mol}}{\mathrm{cm}^3}\right) 3.226$$
$$= 12.9 \times 10^{-12} \frac{\mathrm{mol}}{\mathrm{sec} \cdot \mathrm{cm}^2}$$ eq. AA-4a

and

$$J_o^{\mathrm{K}} = \left(1 \times 10^{-6} \frac{\mathrm{cm}}{\mathrm{sec}}\right) \left(150 \times 10^{-6} \frac{\mathrm{mol}}{\mathrm{cm}^3}\right) 0.1487$$
$$= 22.3 \times 10^{-12} \frac{\mathrm{mol}}{\mathrm{sec} \cdot \mathrm{cm}^2}$$ eq. AA-5a

Thus $J_i^{\mathrm{K}} / J_o^{\mathrm{K}} = 0.578$, which is nearly identical to the value obtained from the flux ratio equation (see Chapter IV on *Diffusion and Permeability*). For Na^+ influx and efflux, assuming a P_{Na} of 2×10^{-8} cm/sec ($P_{\mathrm{Na}}/P_{\mathrm{K}}$ ratio of 0. 02)

$$J_i^{\mathrm{Na}} = -P_{\mathrm{Na}}[\mathrm{Na}^+]_o \frac{VF}{RT} \frac{1}{1 - e^{VF/RT}}$$ eq. AA-6

Specifically

$$J_i^{\mathrm{Na}} = \left(2 \times 10^{-8} \frac{\mathrm{cm}}{\mathrm{sec}}\right) \left(150 \times 10^{-6} \frac{\mathrm{mol}}{\mathrm{cm}^3}\right) 3.226$$
$$= 9.68 \times 10^{-12} \ \mathrm{mol/sec \cdot cm^2}$$ eq. AA-6a

and

$$J_o^{\mathrm{Na}} = P_{\mathrm{Na}}[\mathrm{Na}^+]_i \frac{VF}{RT} \frac{1}{1 - e^{-VF/RT}}$$ eq. AA-7

Specifically

$$J_o^{\mathrm{Na}} = \left(2 \times 10^{-8}\right)\left(15 \times 10^{-6}\right) 0.1487$$
$$= 0.044 \times 10^{-12} \ \mathrm{mol/sec \cdot cm^2}$$ eq. AA-7a

Thus, the net passive Na^+ influx is 9.68 pmol/sec·cm^2 minus 0.04 pmol/sec·cm^2 or 9.64 pmol/sec·cm^2. The ratio of Na^+ influx to Na^+ efflux is 220(9.68/0.044), which is close to the value of 217 calculated from the flux ratio equation (see Chapter IV on *Diffusion and Permeability*). The value for the net Na^+ influx in resting membrane of 9.64 pmol/sec·cm^2 is close to the value for net K^+ efflux of 9.4 pmol/sec·cm^2 (efflux of 22.3 pmol/sec·cm^2 minus influx of 12.9 pmol/sec·cm^2).

Because the net flux of K^+ (J_K) is equal to

$$J_K = J_o^{\,K} - J_i^{\,K} \qquad \text{eq. AA-8}$$

then from equations AA1, 2, 4, and 5:

$$J_K = P_K \frac{VF}{RT} \frac{[K^+]_i}{1 - e^{-VF/RT}} + P_K \frac{VF}{RT} \frac{[K^+]_o}{1 - e^{VF/RT}} \qquad \text{eq. AA-9}$$

From this, the following was derived:

$$J_K = P_K \frac{VF}{RT} \frac{[K^+]_o - [K^+]_i e^{VF/RT}}{1 - e^{VF/RT}} \qquad \text{eq. AA-9a}$$

Now because $J = I/zF$ (see Chapter IV on *Diffusion and Permeability*), substitution into Equation AA-9a and rearrangement gives

$$I_K = P_K V \frac{F^2}{RT} \frac{[K^+]_o - [K^+]_i e^{VF/RT}}{1 - e^{VF/RT}} \qquad \text{eq. AA-10}$$

where I_K is the net K^+ current (outward).

The derivation of the equations, with their assumptions, are given in detail by Goldman (1943) and by Hodgkin and Katz (1949). In brief, the major assumptions are: (a) that the electric field is constant through the thickness of the membrane, i.e., that the voltage gradient has a linear profile; (b) that the membrane is homogeneous; (c) that the ions move under diffusion forces and electric forces, similar to what occurs in free solution; and (d) that the concentrations of ions at the edges of the membrane are directly proportional to those in the bulk solutions bounding the membrane.* Thus, these investigators arrived at equations for K^+, Na^+, and Cl^- currents (K^+ is only presented here as the example)

$$I_K = RTU_K \frac{d[K]}{dx} + [K]U_K F \frac{d\psi}{dx} \qquad \text{eq. AA-11}$$

where $d[K]/dx$ is the concentration gradient of K^+ through the membrane, $d\psi/dx$ is the potential gradient through the membrane, U_K is the mobility of K^+ in the membrane, and I_K is the K^+ current density (constant throughout the membrane). If $d\psi/dx$ is regarded as constant and equal to $-V/\delta$, where V is the PD across the membrane (inside potential minus the outside potential)

$$V \equiv E_m = V_i - V_o$$

and δ is the membrane thickness, integration of Equation AA-11 allowed them to arrive at

---

* Some of these assumptions have been critically examined in several articles, including one by MacGillivray & Hare (1969). They concluded that the thickness of the membrane diffusion barrier must be much less than 100 Å for the constant field assumption to be applicable.

$$I_{\mathrm{K}} = \frac{U_{\mathrm{K}} F V}{\delta} \frac{\beta_{\mathrm{K}}[\mathrm{K}]_o - \beta_{\mathrm{K}}[\mathrm{K}]_i e^{VF/RT}}{1 - e^{VF/RT}} \qquad \text{eq. AA-12}$$

where β_{K} is the partition coefficient for K^+ at the outer and inner edges of the membrane, and hence $\beta_{\mathrm{K}}[\mathrm{K}]_{\mathrm{o}}$ and $\beta_{\mathrm{K}}[\mathrm{K}]_{\mathrm{i}}$ are the concentrations of K^+ at the outer and inner edges of the membrane, respectively

$$\frac{[\mathrm{K}]\text{ inner edge}}{[\mathrm{K}]_{\mathrm{i}}} = \frac{[\mathrm{K}]\text{ outer edge}}{[\mathrm{K}]_{\mathrm{o}}} = \beta_{\mathrm{K}}$$

The permeability coefficient for K^+ (P_K) is defined as

$$P_K = U_K \frac{\beta_K}{\delta} \frac{RT}{F} \qquad \text{eq. AA-13}$$

$$\frac{\mathrm{cm}}{\mathrm{sec}} = \frac{\mathrm{cm/sec}}{\mathrm{V/cm}} \frac{1}{\mathrm{cm}} \mathrm{V}$$

Thus, similar to the modified Fick's diffusion equation discussed in Chapter IV on *Diffusion and Permeability*, in which P was equal to the diffusion coefficient (D) in the membrane divided by the membrane thickness (δ), P in equation AA-13 is proportional to the mobility in the membrane (U) divided by membrane thickness (δ). Equation AA-13 can be rearranged as

$$P_{\mathrm{K}} \frac{F}{RT} = U_{\mathrm{K}} \frac{\beta_{\mathrm{K}}}{\delta} \qquad \text{eq. AA-13a}$$

Substituting this into Equation AA-12 yields

$$I_K = P_K V \frac{F^2}{RT} \frac{[K]_o - [K]_i e^{VF/RT}}{1 - e^{VF/RT}} \qquad \text{eq. AA-14}$$

$$\frac{A}{cm^2} = \frac{cm}{sec} \, V \, \frac{1}{V} \frac{coul}{mol} \left(\frac{mol}{cm^3} - \frac{mol}{cm^3} \right)$$

$$= \frac{coul}{sec} \frac{1}{cm^2}$$

This is the same equation as Equation AA-10. Similar equations were arrived at for the net Na^+ and Cl^- currents

$$I_{Na} = P_{Na} V \frac{F^2}{RT} \frac{[Na^+]_o - [Na^+]_i e^{VF/RT}}{1 - e^{VF/RT}} \qquad \text{eq. AA-15}$$

and

$$I_{Cl} = P_{Cl} V \frac{F^2}{RT} \frac{[Cl^-]_i - [Cl^-]_o e^{VF/RT}}{1 - e^{VF/RT}} \qquad \text{eq. AA-16}$$

From Equations AA-14, AA-15, and AA-16, the total ionic current density ($I = I_K + I_{Na} + I_{Cl}$) through the membrane is therefore given [see Hodgkin & Katz (1949)] by

$$I = P_K V \frac{F^2}{RT} \frac{w - y e^{VF/RT}}{1 - e^{VF/RT}} \qquad \text{eq. AA-17}$$

where

$$w = [K^+]_o + \frac{P_{Na}}{P_K} [Na^+]_o + \frac{P_{Cl}}{P_K} [Cl^-]_i$$

$$y = [K^+]_i + \frac{P_{Na}}{P_K}[Na^+]_i + \frac{P_{Cl}}{P_K}[Cl^-]_o$$

(The reader should carefully note the signs used for the potentials here and elsewhere. The early convention referred to all potentials, i.e., E_m and E_i , outside solution to inside solution, whereas the modern sign convention is inside to outside, representing an external ground.) By designating the PD in the absence of a net ionic current as E ($V = E$ when $I = 0$), Hodgkin and Katz derived from Equation AA-17

$$E = \frac{-RT}{F} \ln \frac{[K^+]_i + \frac{P_{Na}}{P_K}[Na^+]_i + \frac{P_{Cl}}{P_K}[Cl^-]_o}{[K^+]_o + \frac{P_{Na}}{P_K}[Na^+]_o + \frac{P_{Cl}}{P_K}[Cl^-]_i} \qquad \text{eq. AA-18}$$

which is the form of the constant field equation useful for calculating the resting potential as a function of $[K]_o$ for different P_{Na}/P_K ratios (see Chapter VI on *Determination of Resting Potential).*

The membrane conductance (G_m) is defined as the slope through the origin (i.e., at infinitesimally small current) of the steady-state current/voltage plot

$$G_m = \left(\frac{dI}{dV}\right)_{I \to 0} \qquad \text{eq. AA-19}$$

where dV is the voltage derivative. Hodgkin and Katz (1949) derived the following expression from the constant field theory when net $I = 0$ and $V = E$

$$G_m = P_K E \frac{F^3}{R^2T^2} \frac{yw}{y-w} \qquad \text{eq. AA-20}$$

By measuring G_m and the P_{Na}/P_K ratios, one can compute the numerical value of P_K from Equation AA-20. This value of P_K can then be used in Equation AA-14 to obtain I_K. When $I = 0$ and $V = E$, however, a convenient method for obtaining the individual ionic currents is to use the following relations

$$I_K = G_m \frac{RT}{F} \left(\frac{[K^+]_o}{w} - \frac{[K^+]_i}{y} \right) \qquad \text{eq. AA-21}$$

$$I_{Na} = G_m \frac{RT}{F} \frac{P_{Na}}{P_K} \left(\frac{[Na^+]_o}{w} - \frac{[Na^+]_i}{y} \right) \qquad \text{eq. AA-22}$$

$$I_{Cl} = G_m \frac{RT}{F} \frac{P_{Cl}}{P_K} \left(\frac{[Cl^-]_i}{w} - \frac{[Cl^-]_o}{y} \right) \qquad \text{eq. AA-23}$$

From the constant field theory of Goldman, Hodgkin, and Katz, the relation between P_K and g_K, the membrane (chord) conductance for K^+ (in mho/cm^2 or $\Omega^{-1} \cdot \text{cm}^{-2}$) for small K^+ current is:

$$g_K = -P_K\,[K^+]_o\,E \frac{F^3}{R^2T^2} \frac{1}{1-e^{-EF/RT}} \qquad \text{eq. AA-24}$$

The corresponding units are

$$\frac{\Omega^{-1}}{cm^2} = \frac{cm}{\sec}\frac{mol}{cm^3} V \frac{1}{V}\frac{1}{V}\frac{coul}{mol} \text{(dimensionless)}$$
$$= \frac{coul}{\sec}\frac{1}{V}\frac{1}{cm^2}$$
$$= \frac{amp}{V\ cm^2} = \frac{mho}{cm^2} \qquad \text{eq. AA-24a}$$

Conductance (mho) equals current (amp) divided by voltage (V). Hence, conductance is a function of permeability, concentration, and potential. Although changes in P_K give proportional changes in g_K , conductance is related to the ability to carry current, whereas permeability is not.* The comparable equation for g_{Na} is:

$$g_{Na} = P_{Na}\,[Na^+]_o\; E\,\frac{F^3}{R^2T^2}\,\frac{1}{1-e^{EF/RT}} \qquad \text{eq. AA-25}$$

Note that in these equations for g_K and g_{Na} it is assumed that the internal ion concentrations stay constant at their normal levels. Substitution into Equation AA-24, assuming a P_K of 1×10^{-6} cm/sec, gives a calculated g_K of 4.5×10^{-5} mho/cm^2 and R_K of $22 \times 10^3\ \Omega\cdot$cm^2

---

* For example, vertebrate skeletal muscle sarcolemma has a high P_{Cl} and g_{Cl}, but in Cl^--free solution after equilibration and return to the original resting potential, $g_{Cl} = 0$, whereas P_{Cl} stays the same; that is, a change in conductance took place without a change in permeability. Because conductance is defined as the net current flow per unit voltage, if there is no net current flow of an ion under a driving voltage, conductance for that ion is zero, regardless of the P value.

$$g_K = \left(1\times10^{-6}\frac{\text{cm}}{\text{sec}}\right)\left(4\times10^{-6}\frac{\text{mol}}{\text{cm}^3}\right)\left(80\times10^{-3}\text{ V}\right)\left(\frac{10^5\frac{\text{coul}}{\text{mol}}}{6.75\times10^{-4}\text{ V}^2}\right)\left(\frac{1}{0.954}\right)$$

$$= 4.5\times10^{-5}\frac{\text{coul}}{\text{sec}\cdot\text{cm}^2\cdot\text{V}}$$

$$= 4.5\times10^{-5}\frac{\text{mho}}{\text{cm}^2}$$

eq. AA-24a

and

$$R_K = \frac{1}{g_K}$$ eq. AA-24b

$$= \frac{1}{4.5\times10^{-5}\frac{\text{mho}}{\text{cm}^2}}$$

$$= 22.2\times10^3\ \Omega\cdot\text{cm}^2$$

This calculated R_K is close to the value measured for frog sartorius fibers in Cl^--free solution ($g_{Cl} = 0$) (Sperelakis et al., 1967).

From Equations AA-14 and AA-24 it can be seen that the relation between conductance and net flux is

$$g_K = J_K\frac{F^2}{RT}\ \frac{[\text{K}^+]_o}{[\text{K}^+]_i^{VF/RT} - [\text{K}^+]_o}$$ eq. AA-26

If, for example, the net K^+ efflux in resting membrane was 10 pmol/sec·cm^2, then from Equation AA-26, g_K would be

$$g_K = \left(10x10^{-12}\frac{mol}{\sec\cdot cm^2}\right)\left(\frac{10^5\,coul/mol}{26x10^{-3}V}\right)\left(\frac{4\ mM}{150mM\dfrac{1}{e^{+80mV/26mV}} - 4mM}\right)$$

$$= \left(0.385\ x\ 10^{-4}\right)\left(\frac{4}{4-(150)(0.046)}\right)$$

$$= \left(0.385x10^{-4}\right)(1.38) \qquad \text{eq.AA}-26\text{a}$$

$$= 5.3x10^{-5}\ mho/cm^2$$

and

$$R_K = \frac{1}{g_K} \qquad \text{eq. AA-26b}$$

$$= \frac{1}{5.3\times10^{-5}} = 0.188\times10^{-5}$$

$$= 18.8\ \text{K}\Omega\cdot\text{cm}^2$$

Thus, again a reasonable value for K^+ resistance is obtained. Note that the right-hand fraction term in Equation AA-26, containing the concentration and potential terms, yields a number close to one, i.e., the term has only little effect on the calculated value of g_K. As the resting potential is reduced the fraction becomes smaller; hence, the K^+ conductance becomes reduced, thus providing rectification.

Appendix VI-B

DERIVATION OF CHORD CONDUCTANCE EQ.

Ohm's law states that the current (I) is equal to the voltage (E) either divided by the resistance (R) or multiplied by the conductance ($g = 1/R$)

$$I = E/R = gE \qquad \text{eq. AB-1}$$

When dealing with solutions, the voltage or driving force must take into account both the concentration force as well as the electrical force. In this case, the ionic current (I_i) is a product of the conductance for a given ion times the total driving force on that ion (E_m–E_i)

$$I_i = g_i(E_m - E_i) \qquad \text{eq. AB-2}$$

In a resting cell membrane (stable resting potential), the total ionic current must be zero; otherwise the membrane potential would change. Therefore, the K^+ current in an outward direction (I_i) must be equal and opposite to the Na^+ current (I_{Na}) entering the cell (neglecting Ca^{2+}, Cl^-, and minor ions)

$$I_K = -I_{Na} \qquad \text{eq. AB-3}$$

Therefore

$$I_K + I_{Na} = 0 \qquad \text{eq. AB-3a}$$

Substituting the equations for ionic currents from Equation AB-2

$$g_K(E_m - E_K) + g_{Na}(E_m - E_{Na}) = 0 \qquad \text{eq. AB-4}$$

Algebraic manipulations give

$$0 = g_K E_m - g_K E_K + g_{Na} E_m - g_{Na} E_{Na} \qquad \text{eq. AB-4a}$$

$$g_K E_m + g_{Na} E_m = g_K E_K + g_{Na} E_{Na}$$

$$E_m (g_K + g_{Na}) = g_K E_K + g_{Na} E_{Na}$$

Rearrangement gives

$$E_m = \frac{g_K}{g_K + g_{Na}} E_K + \frac{g_{Na}}{g_K + g_{Na}} E_{Na} \qquad \text{eq. AB-5}$$

Equation AB-5 is the **chord conductance equation**. The ratios $g_K/(g_K + g_{Na})$ and $g_{Na}/(g_K + g_{Na})$ are the **fractional conductances** (relative) and are dimensionless.

If Cl^- were to be included, the same steps in the derivation would give the chord conductance equation containing a Cl^- term

$$E_m = \frac{g_K}{\Sigma g} E_K + \frac{g_{Na}}{\Sigma g} E_{Na} + \frac{g_{Cl}}{\Sigma g} E_{Cl} \qquad \text{eq. AB-6}$$

where $\Sigma g = g_K + g_{Na} + g_{Cl}$. It must be emphasized, however, that if Cl^- is passively distributed (in equilibrium at the resting E_m), then Cl^- cannot take part in determining the resting potential (although transient movements of Cl^- can affect the membrane potential when Cl^- is shifted off equilibrium or when the membrane is not at rest).

On the other hand, the Ca^{2+} ion is actively transported and is off equilibrium, so its conductance influences the resting potential. Thus, the chord conductance equation containing the Ca^{2+} term is

$$E_m = \frac{g_{\mathrm{K}}}{\Sigma g} E_{\mathrm{K}} + \frac{g_{\mathrm{Na}}}{\Sigma g} E_{\mathrm{Na}} + \frac{g_{\mathrm{Ca}}}{\Sigma g} E_{\mathrm{Ca}} \qquad \text{eq. AB-7}$$

where $\Sigma g = g_{\mathrm{K}} + g_{\mathrm{Na}} + g_{\mathrm{Ca}}$.

The chord conductance equation, of course, can be written using resistances rather than conductances. For Equation AB-5 using only K^+ and Na^+ terms, substitution of $R = 1/g$ in the equation and algebraic manipulation gives the following equation

$$E_m = \frac{R_{\mathrm{K}}}{R_{\mathrm{K}} + R_{\mathrm{Na}}} E_{\mathrm{Na}} + \frac{R_{\mathrm{Na}}}{R_{\mathrm{K}} + R_{\mathrm{Na}}} E_{\mathrm{K}} \qquad \text{eq. AB-8}$$

Note that the E_{Na} and E_{K} terms are interchanged from the chord conductance equation. This form of the equation might be termed the **chord resistance equation**.

The chord conductance equation applies only to those situations in which the net ionic *current is zero*, such as when the membrane is at rest. This equation is derived simply from Ohm's law, and one advantage it has over the constant field equation is that it can more easily include divalent cations such as Ca^{2+}.

Appendix VI-C

CIRCUIT ANALYSIS FOR CELL MEMBRANE

Using Ohm's and Kirchhoff's laws and logic, it is possible to see why in the nerve or muscle cell the membrane K^+ dominates the resting potential, and Na^+ or Ca^{2+}, or both, dominates the peak of the action potential. The circuit in Figure VI-AC-1 will be used to show that the battery having the lowest resistance in series with it is the battery that is the most expressed across the network. Before analyzing the circuit rigorously, we can consider three conditions and make some qualitative judgments: 1) If the left resistor (R_1) equals the right resistor (R_2) in Figure VI-AC-1 (regardless of their absolute values), the PD across the network is +150 V (upper terminal positive with respect to the lower terminal), i.e., halfway between both batteries, since both should be equally expressed. 2) If R_2 is made infinite (e.g., open circuit in branch 2) and R_1 is finite, then the PD is exactly +100 V, since the left battery in Figure VI-AC-1 (E_1) cannot be expressed at all. 3) If R_1 is much less than R_2, then the PD approaches +100 V, since E_1 is dominant.

The circuit can also be analyzed quantitatively. For example, if $R_1 = 10\ \Omega$ and $R_2 = 990\Omega$, the exact PD may be reasoned from the following analysis. The current (I) has one magnitude, i.e., is constant, throughout this simple closed circuit, but the current flows upward in branch 2 and downward in branch 1. This occurs because the right battery in Figure VI-AC-1 (E_2) is larger than E_1, and so the net driving force for the net current is in the direction as indicated in Figure VI-AC-1. Therefore the voltage drops produced across R_1 and R_2 are in opposite polarities, as shown in Figure VI-AC-1. The voltage drop across R_1 adds to E_1 to make a greater PD

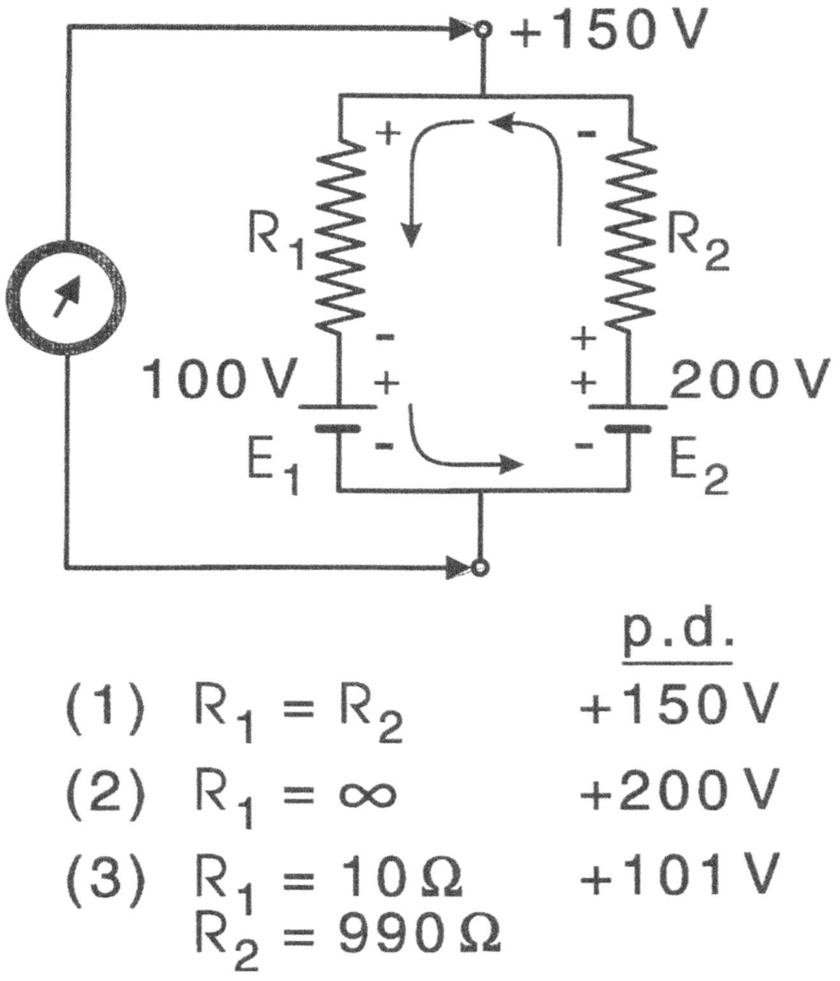

FIGURE VI-AC-1. Circuit diagram of the circuit analysis applicable to the cell membrane, illustrating why the resting potential of a cell is determined by the relative resistances (or conductances). Battery having the lowest resistance in series with it is the battery most expressed across such a network.

across branch 1, like two batteries in series (+ - + -). In contrast, the voltage drop across R_2 subtracts from E_2 to make a smaller PD, like two batteries back to back (- + + -). Therefore the following two equations can be written for the PD across branch 1 [$(PD)_1$] and across branch 2 [$(PD)_2$]

$$(\mathrm{PD})_1 = \mathrm{E}_1 - \mathrm{IR}_1 \qquad \text{eq. AC-1}$$

$$(\mathrm{PD})_2 = \mathrm{E}_2 - \mathrm{IR}_2 \qquad \text{eq. AC-2}$$

To solve these equations we must first calculate the current (I). The net driving force for I is equal to $E_2–E_1$; hence from Ohm's law the net current is equal to ($E_2 - E_1$) divided by the total resistance ($R_1 + R_2$)

$$I = \frac{E_2 - E_1}{R_2 + R_1} \qquad \text{eq. AC-3}$$

$$I = \frac{200\ \mathrm{V} - 100\ \mathrm{V}}{990\ \Omega + 10\ \Omega}$$

$$= \frac{100\ \mathrm{V}}{1{,}000\ \Omega} = 0.1\ \mathrm{A}$$

Now we can enter this value for I into Equation AC-1

$$(PD)_1 = E_1 - IR_1 \qquad \text{eq. AC-1}$$

$$= 100\ \mathrm{V} - (-0.1\ \mathrm{A})(10\ \Omega)$$

$$= 100\ \mathrm{V} - (-1\ \mathrm{V})$$

$$= 100\ \mathrm{V} + 1\ \mathrm{V}$$

$$= 101\ \mathrm{V}$$

The negative sign in the current is because the current in branch 1 produces a voltage drop that adds to E_1. We can also calculate the PD by substituting into Equation AC-2.

$$(PD)_2 = E_2 - IR_2 \qquad \text{eq. AC-2}$$
$$= 200\ \text{V} - (+0.1\ \text{A})(990\ \Omega)$$
$$= 200\ \text{V} - (99\ \text{V})$$
$$= 101\ \text{V}$$

Because the two branches are connected by zero resistances, i.e., they are effectively the same points, $(PD)_1$ must equal $(PD)_2$

$$(PD)_1 = (PD)_2 \qquad \text{eq. AC-4}$$

Thus, the two methods check. To summarize, it is quantitatively demonstrated that the battery with the lowest series resistance is the battery most expressed across this network. This analysis holds true regardless of the absolute values of the resistances or batteries or the polarity of each battery. Other methods of circuit analysis can be used to calculate the PD across such a network, but this method is one of the simplest. The following equation, analogous to the chord conductance equation (see Chapter VI on *Determination of Resting Potential*), can also be derived and used to calculate the PD across the network

$$PD = \frac{R_1}{R_1 + R_2} E_2 + \frac{R_2}{R_1 + R_2} E_1 \qquad \text{eq. AC-5}$$
$$= \frac{10\ \Omega}{10\ \Omega + 990\ \Omega} 200\ \text{V} + \frac{990\ \Omega}{10\ \Omega + 990\ \Omega} 100\ \text{V}$$
$$= \frac{10}{1{,}000} 200\ \text{V} + \frac{990}{1{,}000} 100\ \text{V}$$
$$= 2\ \text{V} + 99\ \text{V}$$
$$= 101\ \text{V}$$

This equation again emphasizes the point that it is the relative resistance that determines which battery is most expressed.

Appendix VI-D

PERMEABILITY COEFFICIENT DETERMINATION

A. Estimation of P_K From Flux Measurements

The first step is to measure the rate coefficient (*k*) for ^{42}K exchange across the cell membrane. To do this, the muscle is loaded with ^{42}K for whatever time is necessary (usually several hours) to bring the specific activity ($^{42}K/K$ total) of the internal K^+ equal to that of the loading solution. Then the muscle is briefly dipped (e.g., for 1 sec) in nonradioactive solution to remove ^{42}K solution adhering to the surface of the muscles. The washout of ^{42}K from the muscle is followed by transferring the muscles through a series of small beakers containing a small volume (e.g., 1 ml) of nonradioactive Ringer's solution; the time interval in each beaker should be short (e.g., 0.5 min) at the beginning of the washout and should be progressively increased to long periods (e.g., 10 min) at the end. The washout should be carried out long enough to wash out most of the ^{42}K from the muscle (e.g., several hours).

Aliquots of each bath are counted for radioactivity, and the total counts in each bath are calculated by taking into account the fractional aliquot that was used. The radioactivity remaining in the muscle is counted by placing the muscle in tissue solubilizer, adding photofluor, and counting. Thus, the total counts present in the muscle at zero time (before the washout began) is obtained by adding the total counts in each washout bath and that remaining in the tissue at the end of the washout. The total counts at zero time is arbitrarily given a value of 100%, and the fractional loss of counts as a function of time is obtained by first subtracting the total counts in each bath from the total counts in the tissue before each soaking began and then dividing by the total counts initially in the tissue. A semilogarithmic plot of these data gives the washout or desaturation curve.

The washout of a single compartment gives a straight line on such a plot, i.e., an exponential decay that represents first-order kinetics. If more than one compartment exists in some series-parallel arrangement, and if these compartments have markedly different rate constants, then the composite washout curve of the tissue exhibits several linear regions of different slopes. Usually there are only two or three such compartments, and for simplicity we assume here that there are only two: a fast interstitial compartment and a slow myoplasmic compartment. By the so-called **curve peeling** (or backward projection) technique, one can obtain approximations of the half-times of washout for each compartment and the relative size of each compartment.* First, the second exponential is extrapolated back to the ordinate; this intercept gives the fractional pool size of the slow (myoplasmic) compartment, and the half-time ($t_{1/2}$) is measured from the graph. [It should be pointed out that the ordinate intercept gives only an approximation of the fractional volume for that compartment (Huxley, 1960), but the error is minimal when the two components have widely separated rate constants; e.g., for a rate constant ratio of 10, the overestimate of the size of the slow compartment is only about 25%.] To obtain the washout curve for the fast (interstitial) compartment, the extrapolated values from the slow compartment are subtracted from the composite curve, and a straight line is fitted to these points.

---

* The curve peeling technique is compatible with the sum of two exponential components, but is not proof of it because the curve could also reflect a continuum of exponential processes (Hill, 1928; Van Liew, 1970; Macdonald et al., 1974).

The intercept of this line with the ordinate gives the relative size of the fast compartment, and the $t_{1/2}$ is measured from the graph.

For small-diameter muscle bundles (e.g., < 1 mm) and for thin, flat muscles (e.g., 0.5 mm thick), the $t_{1/2}$ for the interstitial space is usually less than 1 min, whereas that for the myoplasmic compartment is about 60 min. When using ^{42}K as the tracer, the slow compartment for ^{42}K washout is much larger than the fast compartment, since most of the tissue K^+ is in the myoplasmic compartment. The **time constants (τ)** for each compartment can be measured either directly from the two exponentials on the graph (time for decay to 1/e, or 0.368, or 36.8%) or can be calculated from the $t_{1/2}$

$$\tau = \frac{t_{1/2}}{\ln 2} = \frac{t_{1/2}}{0.693} = 1.44\, t_{1/2} \qquad \text{eq. AD-1}$$

the **rate constant (k)** is obtained by taking the reciprocal of the time constant

$$k = \frac{1}{\tau} \qquad \text{eq. AD-2}$$

The rate constant is usually given in reciprocal minutes (min^{-1}), but for calculation of the fluxes it should be converted to sec^{-1}.

The rate coefficient so determined is then used to calculate the K^+ efflux (J_o^K, in mol/cm^2 per sec) according to the following relationship

$$J_o^{\mathrm{K}} = k[\mathrm{K}]_i \frac{v}{A_s} \qquad \text{eq. AD-3}$$

The corresponding units are

$$\frac{\mathrm{mol}}{\mathrm{sec} \cdot \mathrm{cm}^2} = \left(\mathrm{sec}^{-1}\right)\left(\frac{\mathrm{mol}}{\mathrm{cm}^3}\right)\frac{\mathrm{cm}^3}{\mathrm{cm}^2}$$

where $[\mathrm{K}]_i$ is the intracellular K^+ activity (in mol/cm^3), v is the cell volume (in cm^3), and A_s is the cell surface area (in cm^2) . The v / A_s ratio can be measured by morphometrical or stereological methods (Carmeliet et al., 1976), or idealized cell dimensions and shapes can be assumed. The $[\mathrm{K}]_i$ can be determined from tissue ion analyses (with correction for activity coefficient), from intracellular K^+-selective microelectrodes, or from running an E_m vs log $[\mathrm{K}]_o$ curve and extrapolating to zero potential.

The K^+ permeability coefficient (P_{K}) can then be calculated from the K^+ efflux according to the following equation

$$P_{\mathrm{K}} = J_o^{\mathrm{K}} \frac{RT}{EF} \frac{1-e^{EF/RT}}{[\mathrm{K}]_i e^{EF/RT}} \qquad \text{eq. AD-4}$$

with

$$\frac{\mathrm{cm}}{\mathrm{sec}} = \frac{\mathrm{mol}}{\mathrm{sec} \cdot \mathrm{cm}^2} \frac{\mathrm{V}}{\mathrm{V}} \frac{1}{\mathrm{mol}/\mathrm{cm}^3}$$

where E is the average membrane potential of the cells and must be measured with an intracellular microelectrode in the same Ringer's solution in which the flux studies were carried out. This equation indicates that P_{K} is a direct function of the

K^+ efflux (for a fixed E_m and a fixed $[K]_i$); i.e., if some agent increases K^+ efflux, it must have increased P_K.

By substitution of Equation AD-3 into Equation AD-4, $[K]_i$ cancels out, since it is in both the numerator and denominator, and the useful equation that results is

$$P_K = k\frac{v}{A_s}\frac{RT}{EF}\frac{\left(1-e^{EF/RT}\right)}{e^{EF/RT}} \qquad \text{eq. AD-5}$$

with

$$\frac{\text{cm}}{\text{sec}} = \left(\text{sec}^{-1}\right)\frac{\text{cm}^3}{\text{cm}^2}\frac{\text{V}}{\text{V}}$$

This equation indicates that P_K is a direct function of the rate constant (for a fixed cell geometry and membrane potential). Thus, the permeability coefficient is independent of the ion concentration, as would be expected, because the rate constant is based on a fractional exchange rate (i.e., is independent of the total ions present) and because a first-order rate constant is based on a fractional decay rate. A sample calculation of P_K for chick embryonic hearts (19 days old) bathed in 2.5 mM K^+-Ringer's solution from the work of Carmeliet et al. (1976) follows ($\tau = 48.5$ min, $v/A_s = 1.24 \times 10^{-4}$ cm, $E = -0.085$ V)

$$P_K = (3.44 \times 10^{-4}\,\text{sec}^{-1})(1.24 \times x10^{-4}\,cm)\left(\frac{-0.026V}{-0.085V}\right)\left(\frac{1-e^{(0.026V}}{e^{(-0.085V)(385V^{-1}}}\right)$$

$$= \left(4.266 \times 10^{-8}\,\frac{cm}{\text{sec}}\right)(0.306)\,\frac{1-e^{-3.273}}{e^{-3.273}}$$

$$= \left(1.305 \times x10^{-8}\,\frac{cm}{\text{sec}}\right)\frac{1-0.0379}{0.0379}$$

$$= 3.31 \times 10^{-7}\,\frac{cm}{\text{sec}} \qquad \text{eq. AD} - 5\text{a}$$

This final value is similar to, but not identical to, the value of 3.10×10^{-7} cm/sec (see Table VI-AD-1) calculated by Carmeliet et al. (1976), the small discrepancy perhaps being caused by my having to read the E_m value from their Figure 4. This value of P_K for chick ventricle is close to the value of 1.66×10^{-7} cm/sec obtained for cow Purkinje fibers (Carmeliet & Verdonck, 1977). As demonstrated by the work of Carmeliet and colleagues (Carmeliet et al., 1976; Carmeliet & Verdonck, 1977), there is a substantial decrease in P_K in 0 mM $[K]_o$ in both cow Purkinje fibers and chick hearts (see Table VI-AD-1 and Table VI-AD-2); that is, in cow Purkinje fibers the P_K increases between 0 mM and 5.4 mM $[K]_o$ and in chick hearts P_K increases between 0 mM and 2.5 mM $[K]_o$. One truth that can be gleaned from Equation AD-5 is that if Na^+ influx studies are done on the same tissue under similar conditions (including the same E_m value) and if the comparable equation for P_{Na} is divided into Equation AD-5, then all else cancels out, leaving the ratio of permeabilities equal to the ratio of rate constants

$$\frac{P_K}{P_{Na}} = \frac{k_K}{k_{Na}} \qquad \text{eq. AD-6}$$

TABLE VI-AD-1

Influence of extracellular [K^+] on K^+ permeability of chick embryonic hearts.

	P_K cm/sec (x 10^7)		P_{Na} cm/sec (x 10^7)		$\frac{P_{Na}}{P_K}$	
$[K]_o$	6-8 days	18-20 days	6-8 days	18-20 days	6-8 days	18-20 days
0	0.33	0.95	0.044	0.063	0.133	0.066
2.5	1.32	3.10	0.042	0.053	0.032	0.017
5.0	1.32	2.40	0.057	0.046	0.043	0.019
10	1.44	2.42				
20	1.39	2.65				

Values for P_{Na} were calculated from measured P_K values (from flux studies) and measured P_{Na}/P_K ratios (obtained from the constant field equation). The flux density values, on which the P_K calculations are based, were obtained from measured volume/surface area ratios (by a stereologic method) of 1.24×10^{-4} cm for both 7-day-old and 19-day-old hearts. The measured [K^+] values were 151 mM (6-8 days) and 122 mM (18-20 days), and the corresponding $[Na^+]_i$ values were 16 mM and 15 mM. [Data from Carmeliet et al. 1976.]

TABLE VI-AD-2

Influence of $[K]_o$ and $[Cl]_o$ on P_K of cow Purkinje fibers.

		P_K cm/sec (x 10^7)		P_{Cl} cm/sec (x 10^7)	$\frac{P_{Cl}}{P_K}$
$[K]_o$ mM	E_m mV	in Cl^-	Cl^--free		
0	-45	0.34	0.23	0.59	1.724
5.4	-75	1.66	1.09	0.34	0.205
16.2	-60	1.30	0.87	0.58	0.446
27	-45	1.08	0.75	—	—
54	-30	1.02	0.66	—	—

Data from Carmeliet and Verdonck, 1977.

The P_{Cl} can also be measured from flux studies. For example, Carmeliet and Verdonck (1977) measured Cl^- efflux in cow Purkinje fibers using ^{36}Cl. In 5.4 mM $[K]_o$-Ringer's solution the Cl^- rate constant (k_{Cl}) was 40.0 × 10^{-5} (compared to k_K of 11.7 × 10^{-5} sec^{-1}); J_o^{Cl} was 3.69 × 10-12 mol/cm^2 per sec (compared to 4.78 × 10^{-12} for J_o^K); and the calculated P_{Cl} was 0.34 × 10^{-7} cm/sec (compared to P_K of 1.66 × 10^{-7}). Thus the P_{Cl}/P_K ratio was 0.205. As shown in Table VI-AD-2, however, P_{Cl} seemed to vary somewhat as a function of $[K]_o$. In addition, P_K was affected by the absence of Cl^-, as shown in Table VI-AD-2, P_K decreased substantially in Cl^--free solution. Hence, Carmeliet and Verdonck concluded that the increase in membrane resistance measured in Cl^--free solution is partially due to a decrease in K^+ conductance, as well as to a decrease in Cl^- conductance to zero.

B. Estimation of the P_{Na}/P_K Ratio From Membrane Potential Measurements

The P_{Na}/P_K ratio can be obtained by running an E_m vs. log $[K]_o$ curve, as was illustrated in Figure VI-2 (and will be illustrated in Chapter X (Fig. X-1)) and described in Chapter VI *Determination of Resting Potential*, sections B and D. Running the entire curve has the advantage of allowing a fairly accurate measurement of $[K]_i$ (by extrapolation of the curve to zero potential), thus obviating the necessity of a chemical ion analysis or use of K^+-selective microelectrodes. In addition, the P_{Na}/P_K ratio can be calculated for several $[K]_o$ values, and hence the effect of $[K]_o$ on the ratio can be determined. It should be pointed out that attempts to determine the P_{Na}/P_K ratio at $[K]_o$ levels greater than 20 mM are less and less accurate, because as shown in Figures VI-2 and X-1, the curves for different P_{Na}/P_K ratios converge toward a common point; therefore a small error of one or two millivolts in the measurement of E_m can cause a large error in the calculation of the P_{Na}/P_K ratio. Another way of stating this is that at a high $[K]_o$ the E_m is relatively insensitive to the P_{Na}/P_K ratio.

The Goldman constant field equation (given in Chapter VI on *Determination of Resting Potential*) is used in the calculation of the P_{Na}/P_K ratio

$$E_m = -60 \text{ mV} \log \frac{[K]_i + \frac{P_{Na}}{P_K}[Na]_i}{[K]_o + \frac{P_{Na}}{P_K}[Na]_o} \qquad \text{eq. AD-7}$$

Because E_m and $[K]_i$ are measured and because $[K]_o$ and $[Na]_o$ are known, this equation can be solved for the P_{Na}/P_K ratio if $[Na]_i$ is either measured or assumed. Algebraic rearrangement of the constant field equation gives

$$\frac{P_{Na}}{P_K} = \frac{[K]_i - [K]_o\left(\text{antilog}\frac{E_m}{-60 \text{ mV}}\right)}{[Na]_o\left(\text{antilog}\frac{E_m}{-60 \text{ mV}}\right) - [Na]_i} \qquad \text{eq. AD-7a}$$

As an illustration, if $[K]_i$ is measured to be 150 mM, $[K]_o$ is 4 mM, $[Na]_o$ is 148 mM, $[Na]_i$ is assumed to be 15 mM, and E_m is measured to be -80 mV, then

$$\frac{P_{Na}}{P_K} = \frac{(150 \text{ mM}) - \left[(4 \text{ mM})\left(\text{antilog}\frac{-80 \text{ mV}}{-60 \text{ mV}}\right)\right]}{(148 \text{ mM})\left(\text{antilog}\frac{-80 \text{ mV}}{-60 \text{ mV}}\right) - (15 \text{ mM})} \qquad \text{eq AD-7b}$$

$$= \frac{150 - 4(21.6)}{148(21.6) - 15}$$

$$= \frac{63.6}{3182} = 0.020$$

If $[Na]_i$ is assumed to be 1.5 mM, then the P_{Na}/P_K ratio calculates to be 0.020 also, thus demonstrating that the calculated P_{Na}/P_K ratio is insensitive to $[Na]_i$.

In this method of measuring P_{Na}/P_K , it is assumed that Cl^- is passively distributed and does not influence the resting potential under steady-state conditions. If Cl^- is not passively distributed, then these measurements must be done in Cl^--free solution (Cl^- substituted by an impermeant anion), so that E_m reflects only Na^+ and K^+. It is also assumed that an electrogenic pump potential contribution to E_m is negligible; if not, the measurements must be done rapidly in the presence of a pump-blocking concentration of ouabain.

C. Calculations of P_{Na} From P_K and P_{Na}/P_K

The P_{Na} can either be measured from Na^+ influx studies in a manner similar to that for P_K , or P_{Na} can simply be calculated from the measured P_K value and from the measured P_{Na}/P_K ratio. For the example given in the preceding subsection ($P_{Na}/P_K = 0.020$, and $P_K = 3.31 \times 10^{-7}$ cm/sec)

$$P_{Na} = (P_K)\left(\frac{P_{Na}}{P_K}\right) \qquad \text{eq. AD-8}$$

$$= (3.31 \times 10^{-7}\ \text{cm/sec})\ (0.020)$$

$$= 0.0662 \times 10^{-7}\ \text{cm/sec}$$

Carmeliet et al. (1976) used this method to calculate a P_{Na} of 0.53×10^{-8} cm/sec for 19-day-old chick embryonic hearts (in 2.5 mM $[K]_o$); see Table VI-AD-1.

Appendix VI-E

HALF-CELL POTENTIALS

The transmembrane resting potential recorded by an intracellular microelectrode is not to be confused with electrode half-cell potentials. For example, if two beakers containing NaCl at 0.1 M and at 0.01 M were joined by a salt bridge (agar-NaCI), and if a Ag-AgCl half-cell electrode were placed in each beaker, then a PD of 59 mV would be recorded between the two electrodes, because the potential of each half-cell, reversible to Cl^- ions, would be different (Fig. VI-AE-1A). In this example, the beaker containing the higher Cl^- concentration would be negative, and the one with the lower Cl^- concentration would be positive. The AgCl coat of the electrode immersed in the lower Cl^- concentration would have the greater tendency to solubilize and ionize, leaving this electrode positive; conversely, the AgCl coat of the electrode immersed in the higher Cl^- concentration would have the lesser tendency to solubilize and actually would tend to deposit more AgCl, stealing a positive charge from the wire and thus leaving that electrode negatively charged.

A positive potential is applied to electroplate the Ag wires with AgCl by electrophoresing Cl^- to the Ag wires, as shown in Fig. VI-AE-1B. Note that the resting potential recorded in biological cells is not a function of the half-cell potentials (i.e., an artifact), because the solutions bathing the half-cells (e.g., Ag-AgCl wires or calomel half-cells) remain constant; i.e., the half-cell potentials stay the same whether the microelectrode is inside or outside the cell. The two half-cell potentials are nearly equal in magnitude and so cancel each other. Any small amount of difference between the two half-cell potentials (e.g., a few mV) when the two electrodes are in

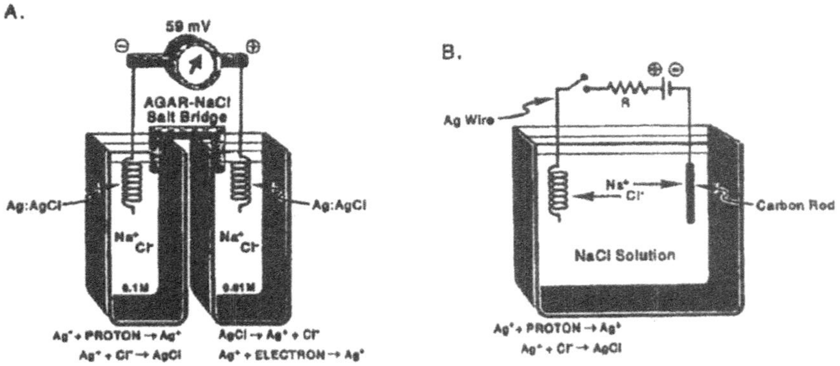

FIGURE VI-AE-1. Half-cell electrode potentials: **A:** Two Ag-AgCl half-cell electrodes are bathed in two different solutions containing Cl^- ion at different concentrations (0.1 M and 0.01 M, in example). The Ag-AgCl half-cells are reversible to Cl^-, and therefore the half-cell potential depends on the Cl^- concentration in which the electrode is bathed. Electrode bathed in the highest Cl^- concentration has a more negative value than the electrode in lower Cl^- concentration. Thus the two half-cell potentials are not equal to one another, and their potentials do not cancel out, as is normally true. There is a net PD produced by the two unequal half-cell potentials, the electrode in the higher Cl^- concentration being negative and the electrode in the lower Cl^- concentration being positive. The two solutions are connected by an agar salt bridge to complete the circuit. **B:** For electroplating a silver wire electrode with AgCl, the silver wire must be made positive so that Cl^- ions can be electrophoresed through the solution to react with silver atoms to plate AgCl.

the same Ringer's solution is arbitrarily called the zero potential (in practice, with the microelectrode in position, any small microelectrode tip potential, e.g., up to 5 mV, would be included in the zeroing procedure). The cell's resting potential is added in series with the half-cell potential, and thus the recording system gives the true transmembrane resting potential.

Chapter VII

Electrochemical Driving Forces and Membrane Ionic Currents

A. Current Equations

In any electrolytic system the force tending to drive net movement of a charged particle must take into account both the concentration (or chemical) force and the electrical force. Because the electrical force is the PD across the cell membrane (E_m), the net driving force for any ionic species then is equal to (E_m–E_i). The driving forces for Na^+, Ca^{2+}, and K^+ are illustrated in Figure VII-1 for a cardiac muscle cell having a resting potential of -80 mV. At 37° C, E_{Na} is about +60 mV, because the ratio of concentrations is about 10; the battery is positive (i.e., the inside plate is positive with respect to the outside plate), because the higher concentration is outside. The E_{Ca} is about +130 mV, because the ratio of concentrations of free ions is about 20,000; the Ca^{2+} battery is directed in the same direction as the Na^+ battery. The E_K is about -94 mV, because the ratio of concentrations is about 37; the battery polarity is reversed from that for Na^+ and Ca^{2+}, because the concentration gradient is in the opposite direction.

The total force tending to drive Na^+ from outside to inside the cell is the algebraic difference of the membrane potential (E_m of -80 mV) and the Na^+ equilibrium potential (E_{Na} of +60 mV), or -140 mV. (All potentials are referred to the extracellular fluid as the ground or reference potential.) The negative sign represents inwardly directed positive current (I_{Na}). Similarly, the total force tending to drive Ca^{2+} into the cell (E_m–E_{Ca}) is -210 mV [-80 mV – +130 mV)]. The total force tending to drive K^+ out of the cell (E_m–E_K) is +14 mV [-

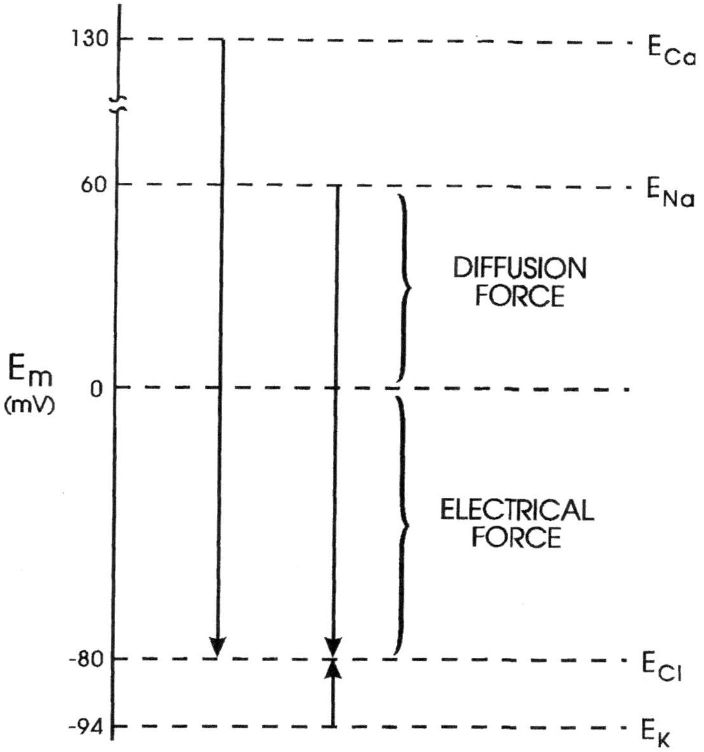

FIGURE VII-1. Representation of the electrochemical driving forces for Na^+, Ca^{2+}, K^+, and Cl^-. Equilibrium potentials for each ion (e.g., E_{Na}) are positioned vertically according to their magnitude and sign; they were calculated from Nernst equation for given set of extracellular and intracellular ion concentrations. Measured resting potential is assumed to be -80 mV. Electrochemical driving force for an ion is the difference between its equilibrium potential (E_i) and the membrane potential (E_m), that is, ($E_i - E_m$). Thus at rest the driving force for Na^+ is the difference between E_{Na} and the resting E_m; if E_{Na} is +60 mV and resting E_m is -80 mV, the driving force is 140 mV; that is, the driving force is the algebraic sum of the diffusion force and the electrical force, and is represented by the length of the arrows in the diagram. Driving force for Ca^{2+} (about 210 mV) is even greater than that for Na^+, whereas that for K^+ is much less (about 14 mV). Direction of the arrows indicates the direction of the net electrochemical driving force, namely, the direction for K^+ is outward, whereas that for Na^+ and Ca^{2+} is inward. If Cl^- is passively distributed, then its distribution across the cell membrane can only be determined by the net membrane potential; for a cell sitting a long time at rest, $E_{Cl} = E_m$ and there is no net driving force.

80 mV – (-94 mV)]; the positive sign indicates outwardly directed positive current (I_K). The driving forces are proportional to the length of the arrows in Figure VII-1; the driving force for Na^+ and Ca^{2+} is the sum of the concentration gradient (E_{Na} or E_{Ca}) and the electrical gradient of -80 mV.

If Cl^- is not actively transported in cardiac muscle, Cl^- distributes itself passively according to the membrane potential. The negatively charged cell interior drives Cl^- out until the force of the concentration gradient built up (E_{Cl}) is equal and opposite to the electrical gradient. Thus at equilibrium E_{Cl} = E_m; there is no driving force and therefore no net Cl^- current (that is, Cl^- influx equals Cl^- efflux). If so, then Cl^- ion cannot participate in the electrogenesis of the resting potential. Some evidence has been presented, however, indicating that in frog heart Cl^- is pumped inward, so that E_{Cl} is considerably less negative (about -40 mV) than the resting potential (Ladle & Walker, 1975).

The current for each ion (I_i) can be calculated from Ohm's law as a product of the ionic conductance (g_i) times the driving force (V)

$$I_i = \frac{V}{R} = g_i V \qquad \text{eq. 1}$$

where R is the resistance. But because in solution the driving force must take into account both electrical and diffusion forces, the actual driving force is (E_m–E_i). Then

$$I_i = g_i (E_m - E_i) \qquad \text{eq. 2}$$

or specifically

$$I_{Na} = g_{Na} (E_m - E_{Na}) \qquad \text{eq. 2a}$$

and

$$I_K = g_K (E_m - E_K) \qquad \text{eq. 2b}$$

If Cl^-, Ca^{2+}, and minor ions can be neglected, then at rest the K^+ current must be equal and opposite to Na^+ current in order to maintain a steady resting potential

$$I_K = -I_{Na} \qquad \text{eq. 3}$$

Thus, although in the resting membrane g_K is much larger than g_{Na}, the driving force for Na^+ is much greater than that for K^+, so that the currents are equal. Hence, there are equally great and continual so-called leak currents of inward Na^+ and outward K^+ down their electrochemical gradients in resting muscle. If active transport were to be blocked, then the system would run down.

The membrane potential of the resting membrane is close to E_K because g_K is greater than g_{Na}, whereas during the rising phase of the action potential there is a large increase in g_{Na}, so that the membrane potential becomes dominated by E_{Na} (i.e., E_m approaches E_{Na}). The relative conductances in the resting membrane can be obtained by substituting Equations 2a and 2b into Equation 3

$$g_K (E_m - E_K) = g_{Na} (E_m - E_{Na}) \qquad \text{eq. 3a}$$

Therefore

$$\frac{g_K}{g_{Na}} = \frac{(E_m - E_{Na})}{(E_m - E_K)} = \frac{140 \text{ mV}}{14 \text{ mV}} = \frac{10}{1} \qquad \text{eq. 3b}$$

Thus, the ratio of conductances can be obtained from the ratio of driving forces.

B. Automaticity

During the pacemaker potential in cells that exhibit them, Equation 3 does not hold. Instead, the inward cation current (Na^+ plus Ca^{2+}) exceeds the outward current (K^+), and depolarization (the pacemaker potential) results: $(I_{Na} + I_{Ca}) > I_K$. When the threshold potential is reached, the cell fires an action potential, and the entire cycle is repeated. During automaticity, the following points need be made: (a) The outward current is less than the inward current, because g_K is decreasing during the time course of the pacemaker potential, and hence decreasing I_K; thus the background inward Na^+ plus Ca^{2+} current could remain constant. (b) The membrane resistivity (R_m) increases during the pacemaker potential, as expected from a decrease in g_K and from anomalous rectification. (c) Many pacemaker cells (especially nodal cells of the heart) have a low g_K, thereby giving a high ratio of g_{Na} / g_K and a low resting potential. (d) The g_{Cl} is generally low in pacemaker cells, so that the Cl^- battery does not act to clamp or hold the membrane potential at the so-called resting level (i.e., Cl^- must not short-circuit the pacemaker mechanism). (e) After the action potential, pacemaker cells almost always have a large hyperpolarizing afterpotential that leads into the pacemaker potential; a change in slope often marks the end of the afterpotential and the beginning of the pacemaker potential. The afterpotential is produced by a large increase in g_K; the subsequent decay in g_K terminates the afterpotential.

(f) In cardiac nodal cells the parasympathetic neurotransmitter (acetylcholine) increases g_K rather selectively, and thereby hyperpolarizes toward E_K and depresses the slope of the pacemaker potential. (g) The sympathetic neurotransmitter (norepinephrine) decreases g_K to some extent, alters the kinetics of the voltage-sensitive K^+ channels, and increases the number of slow Ca^{2+} channels (and hence increases g_{Ca}); the net effect is an increase in the background inward current and a decrease in the outward K^+ current, an effect resulting in a greater rate of spontaneous depolarization. Thus any agent that increases g_K , such as acetylcholine or elevated $[K]_o$, suppresses automaticity, whereas any agent that decreases g_K , such as norepinephrine or Ba^{2+} ion (Sperelakis & Tarr, 1965), enhances automaticity. Finally, any agent that increases the background inward Ca^{2+} (or Na^{+}) current (such as norepinephrine) should enhance automaticity, whereas any agent that decreases the background current (such as verapamil, D-600, Mn^{2+}, or La^{3+}) should suppress automaticity.

C. Electrical Equivalent Circuit of Cell Membrane

The electrical equivalent circuit of the resting cell membrane is given in Figure VII-2. This equivalent circuit is modeled after the Hodgkin-Huxley electrical circuit for nerve membrane (Hodgkin & Huxley, 1952a; Hodgkin, 1964). As illustrated in Figure VII-2A, the outside longitudinal resistance (r_o) is generally low, compared to the internal longitudinal resistance (r_i), normalized for unit length, and therefore for simplicity is represented by a straight line (zero resistance). A low r_o is especially true for a single fiber bathed in a large volume, or for the surface fibers of a nerve or muscle bundle. The Appendix to Chapter I, discusses cables and the various units describing cable properties. The calculated r_i for a single chain of myocardial cells is about 1×10^8 Ω/cm, assuming a myoplasmic resistivity (R_i) of 200 Ω·cm (the myoplasmic resistivity of skeletal muscle fibers, in which it is possible to measure with relative accuracy, is about 200 Ω·

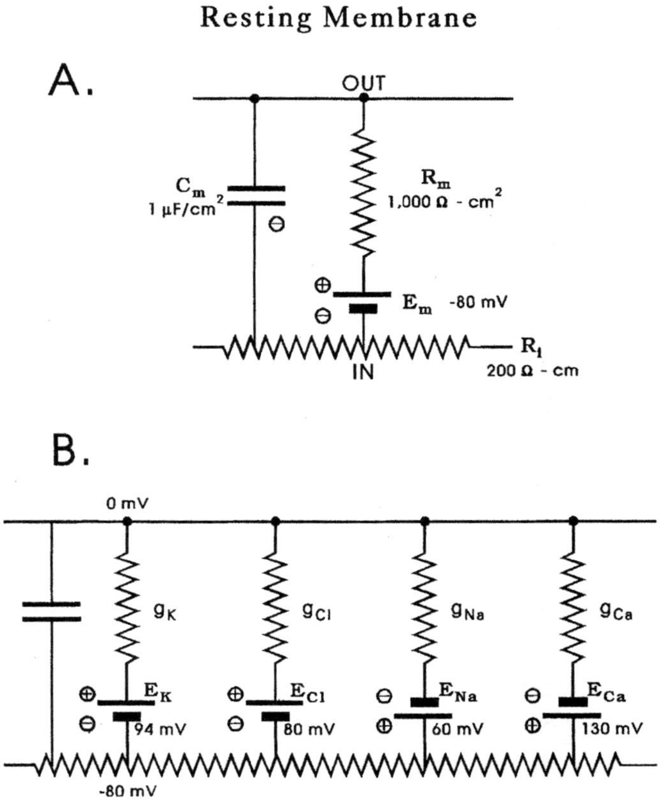

FIGURE VII-2. Electrical equivalent circuits for a cell membrane at rest. **A:** Membrane as a parallel resistance-capacitance circuit, the membrane resistance (R_m) being in parallel with the membrane capacitance (C_m). Resting potential (E_m) is represented by an 80 mV battery in series with the membrane resistance, the negative pole facing inward. **B:** Membrane resistance is divided into its four component parts, one for each of the four major ions of importance: K^+, Cl^-, Na^+, and Ca^{2+}. Resistances for these ions (R_K, R_{Cl}, R_{Na}, and R_{Ca}) are parallel to one another, and represent totally separate and independent pathways for permeation of each ion through the resting membrane. These ion resistances are depicted as their reciprocals, namely, ion conductances (g_K, g_{Cl}, g_{Na}, and g_{Ca}). Equilibrium potential for each ion (e.g., E_K), determined solely by the ion distribution in the steady state and calculated from the Nernst equation, is shown in series with the conductance path for that ion. Resting potential of -80 mV is determined by the equilibrium potentials and by the relative conductances.

cm at 37° C) and assuming cylindrical cells with a radius of 8 μm

$$r_i = \frac{R_i}{\pi a^2} \qquad \text{eq. 4}$$

$$= \frac{200\ \Omega \cdot \text{cm}}{\pi(8 \times 10^{-4})^2}$$

$$\cong \frac{200\ \Omega \cdot \text{cm}}{201 \times 10^{-8}\ \text{cm}^2}$$

$$= 1 \times 10^8\ \Omega/\text{cm}$$

The absolute longitudinal resistance of the myoplasm of one myocardial cell, assuming a cell length (L) of 150 μm, then would be 1.5 MΩ

$$r = R_i \frac{L}{\pi a^2} \qquad \text{eq. 5}$$

$$= \frac{(200\ \Omega \cdot \text{cm})(150 \times 10^{-4}\ \text{cm})}{\pi(8 \times 10^{-4}\ \text{cm})^2}$$

$$\cong 150 \times 10^4\ \Omega$$

$$= 1.5\ \text{M}\Omega$$

where r is the absolute resistance (independent of size). The measured input resistance (i.e., the resistance that an intracellular microelectrode looks into) of a myocardial cell is about 4.5 MΩ (Sperelakis & Shigenobu, 1972). Of this, a maximum of about 0.3 MΩ (0.19–0.38) is due to the myoplasmic resistance, and the remainder of about 4.2 MΩ is due to the cell membrane resistance. If the cell surface area (excluding any T system) is about 75×10^{-6} cm^2

$$\text{Surface area} = \text{circumference} \times \text{length} = 2\pi a\,L$$
$$= 2\pi(8\times10^{-4}\ \text{cm})(150\times10^{-4}\ \text{cm})$$
$$\cong 75\times10^{-6}\ \text{cm}^2$$

then the membrane resistivity (R_m), assuming no electrical coupling between cells is:

$$R_m = R_{cell} \times A_s \qquad \text{eq. 6}$$
$$R_m = (4.2\times10^6\ \Omega)(75\times10^{-6}\ \text{cm}^2)$$
$$= 315\ \Omega\cdot\text{cm}^2$$

If there were substantial electrical coupling, or if there were a T-tubular system, or both, then the resistivity of a patch of membrane would be higher (because of the greater number of parallel resistance pathways). Some estimated values for R_m of myocardial cells are as high as 10,000 $\Omega\cdot\text{cm}^2$.

The specific membrane capacitance (C_m) for myocardial cells generally obtained is about 1–2 $\mu\text{F/cm}^2$ (Fozzard, 1972). In some instances, much higher values have been reported, but these may be erroneous. Giant axons of the squid and other species have C_m values of about 0.7–1.5 $\mu\text{F/cm}^2$. The input capacitance (c_{in}) of one myocardial cell is about 75 pF:

$$c_{in} = C_m \times \text{surface area} \qquad \text{eq. 7}$$
$$= \left(1\times10^{6}\ \text{F/cm}^2\right)\left(75\times10^{-6}\ \text{cm}^2\right)$$
$$= 75\times10^{-12}\ \text{F}$$

See the Appendix to Chapter I for a discussion of capacitance.

In Figure VII-2B, the electrical equivalent circuit for the resting membrane has been expanded to indicate the conductances and the equilibrium potentials (or batteries) for each of the ions of interest, namely, K^+, Cl^-, Na^+, and Ca^{2+}. These are the only ions — with membrane conductances great enough and batteries high enough — whose movements through the resting or excited membrane can significantly alter the membrane potential. As discussed in Chapter V on *Concentration Cells and Equilibrium Potentials*, equilibrium potentials are calculated from the Nernst equation, based on the steady-state distributions for each ion. (Although concentration is used here for convenience, keep in mind that we really mean *activity*, i.e., the free ion concentration multiplied by its **activity coefficient** in that solution; but if the activity coefficients are the same in the external and internal solutions, then they cancel out, and the calculated equilibrium potentials are accurate.) Because the positively charged K^+ ion is much greater in concentration inside the cell, its diffusion down its concentration gradient makes the inside become negative and the outside positive; this polarity is indicated in the equivalent circuit (Fig. VII-2B). If $[K]_i$ is 150 mM and $[K]_o$ is 4.5 mM, then at 37° C, E_K is about -93 mV.

$$E_K = -61 mV \log \frac{[K]_i}{[K]_o} \qquad \text{eq. 8}$$

$$\begin{aligned} E_K &= -61 \text{ mV} \log \frac{150 \text{ mM}}{4.5 \text{ mM}} \\ &= (-61 \text{ mV})(\log 33.3) \\ &= (-61)(1.52) \\ &= -92.9 \text{ mV} \end{aligned}$$

If Cl^- is passively distributed, Cl^- is lowered inside the cell to about 5.0 mM, because the negative resting potential (generally about -80 mV) repels the negatively charged Cl^- ion.

$$E_{Cl} = +61\,mV\ \log \frac{[Cl]_o}{[Cl]_i} \qquad \text{eq. 9}$$

$$= +61\ \text{mV}\ \log \frac{100\ \text{mM}}{5.0\ \text{mM}}$$

$$= -79.4\ \text{mV}$$

Thus, E_{Cl} should be exactly equal to the resting potential if the cell is quiescent for a long period. If the cell is firing action potentials, then E_{Cl} tends to oscillate in value; the mean E_{Cl} should reflect the mean membrane potential (see Chapter II on *Ion Distributions and Na^+-K^+ Pump*). In frog myocardial cells, however, it has been reported that the internal Cl^- activity is independent of whether the tissue is beating or nonbeating (Ladle & Walker, 1975). In quiescent sheep Purkinje fibers, it has been reported from measurements with Cl^--sensitive microelectrodes that E_{Cl} is about -52 mV (mean internal Cl^- activity of 14 mM), compared to a mean E_m of -74 mV (Vaughan-Jones, 1977).

Because the positively charged Na^+ ion is much less inside the cell than outside, diffusion down its concentration gradient makes the inside become positive and the outside negative; this polarity is indicated in the equivalent circuit (Fig. VII-2B). If $[Na]_o$ is 145 mM and $[Na]_i$ is 15 mM (although $[Na]_i$ may actually be considerably less, e.g., 1–6 mM), then at 37° C, E_{Na} is about +60 mV

$$E_{Na} = -61 mV\ \log \frac{[Na]_i}{[Na]_o} \qquad \text{eq. 10}$$

$$E_{Na} = -61\ \text{mV} \log \frac{15\ \text{mM}}{145\ \text{mM}}$$

$$= (+61)\ (\log 9.67)$$

$$= +60.1\ \text{mV}$$

The concentration gradient, and hence equilibrium potential, for Ca^{2+} is in the same direction as that for Na^+. If $[Ca]_o$ is 2 mM and $[Ca]_i$ is 1×10^{-7} M, then at 37° C, E_{Ca} is about +130 mV:

$$E_{Cl} = \frac{-61\ mV}{2} \log \frac{[Ca]_i}{[Ca]_o} \qquad \text{eq. 11}$$

$$E_{Ca} = \frac{-61\ \text{mV}}{2} \log \frac{1 \times 10^{-7}\ \text{M}}{2 \times 10^{-3}\ \text{M}}$$

$$= (+30.5)\,(\log 2 \times 10^{4}) = (+30.5)\,(4.30)$$

$$= +131\ \text{mV}$$

In the equivalent circuit, the overall membrane resistivity (R_m) is broken up into its several component parts (Fig. VII-2B): the resistivity of K^+, Cl^-, Na^+, and Ca^+ (R_K , R_{Cl} , R_{Na} , and R_{Ca} , respectively); that is, each ion has its own diffusion pathway through the membrane. The reciprocal of the resistance (R, in Ω) is the conductance (g, in mho): $g = 1/R$ (see Appendix to Chapter I). The conductance for each ion, in series with the battery for that ion, is shown in Figure VII-2B: g_K , g_{Cl} , g_{Na} , and g_{Ca}. If Cl^- is passively distributed, that ion cannot determine the resting potential. Therefore, the resting potential is determined by the relative conductances for K^+ on the one hand, and for Na^+ and Ca^{2+} on the other hand. In the resting membrane, g_K is much greater than g_{Na} or g_{Ca} , and hence the membrane potential is close to E_K.

In the equivalent circuit for an excited membrane, one main difference from the resting equivalent circuit is that all of the ionic resistances, except that for Cl^-, have arrows drawn through them to represent that the value of the resistance changes as a function of E_m; that is, the conductances are voltage dependent. A second difference is that a given ion, such as K^+ or Na^+, may pass through more than one type of channel. For all channel pathways, except for one K^+ pathway (I_{K1}) and one mixed conductance (Na^+, K^+) pathway (I_f), depolarization increases the ionic conductance. For the anomalous rectifier K^+ pathway (I_{K1}), depolarization decreases this conductance very quickly, and hyperpolarization increases the conductance. This behavior is the so-called K^+ **anomalous rectification**, anomalous in the sense that depolarization decreases its conductance rather than increasing it, as happens in the delayed rectifier K^+ pathway. In a rectifier, current flows more easily in one direction. The delayed rectifier K^+ pathway is also known as **outward-going rectification**, this pathway allowing K^+ efflux more readily than influx. The I_{Kl} pathway is also known as **inward-going rectification**, allowing K^+ to pass more readily inward than outward. For more details, the reader is referred to the book by Noble (1975).

There is no evidence in working myocardial cells that g_{Cl} increases with depolarization, although there is some evidence for this in Purkinje fibers (Dudel et al., 1967). But even without a voltage-dependent conductance change, during the action potential plateau there is a net driving force for inward Cl^- flux, and this Cl^- influx (outward current) acts to hasten repolarization, because the Cl^- current brings negative charge inside the cell.

There is a second type of Na^+ channel in myocardial cells known as the slow Na^+ channel. As the name implies, these channels are kinetically slow, compared to fast Na^+ channels. The slow channels are activated slowly, inactivated slowly, and recover slowly (Rougier et al., 1969); that is, their activation and inactivation gates move more slowly. The slow channels also have a lower (less negative) activation potential (e.g., -35 mV) and a lower inactivation potential (e.g., -25 mV) than the fast Na^+ channels. The slow channels are not blocked by tetrodotoxin (TTX) or saxitoxin (STX), as are the fast Na^+

channels. But the slow channels are blocked by verapamil and D-600, whereas the fast Na^+ channels are relatively unaffected. Lithium ions (Li^+) cannot pass through slow Na^+ channels, as they can in fast Na^+ channels. Young embryonic hearts in the tubular stage do not possess fast Na^+ channels; instead they fire action potentials whose inward current is carried solely by slow channels (see Chapter X on *Developmental Changes in K^+ Permeability*). Smooth muscles never develop fast Na^+ channels, the inward current for their action potentials being carried through slow Na^+ and slow Ca^{2+} channels.

All nerve and muscle membranes probably possess slow Ca^{2+} channels that are voltage activable. These slow Ca^{2+} channels have properties similar to the slow Na^+ channels described in the preceding paragraph. In addition to verapamil and D-600, however, the Ca^{2+} slow channels are selectively blocked by Mn^{2+}(1 mM) and La^{3+} (< 1 mM). Also the Ca^{2+} slow channels in cardiac muscle are dependent on metabolic energy to maintain them in a state of availability for voltage activation; that is, the Ca^{2+} slow channels are blocked by metabolic inhibition at a time when the fast Na^+ channels and K^+ channels are unaffected (Schneider & Sperelakis, 1974). The level of cyclic AMP is often related to the number of slow Ca^{2+} channels available (Shigenobu & Sperelakis, 1972; Tsien et al., 1972; Schneider & Sperelakis, 1974), and it has been proposed that phosphorylation of a protein constituent of the slow Ca^{2+} channel makes the channel available for voltage activation (Schneider & Sperelakis, 1974; Sperelakis & Schneider, 1976). In cardiac and smooth muscles, the inward Ca^{2+} current participates in excitation-contraction coupling. Slow Ca^{2+} channels may also be present in the sarcoplasmic reticulum (SR) membrane of skeletal muscle (see Sperelakis et al., 1973). In nerve an inward Ca^{2+} current takes part in the release of a neurotransmitter from the nerve terminal (excitation-secretion coupling), and in nerve and cardiac muscle the Ca^{2+} influx may lead to K^+ activation (Meech, 1974; Isenberg, 1975).

Separate conductance pathways for each major ion are indicated by the following evidence: (a) The Japanese puffer fish poison, tetrodotoxin (TTX), blocks fast Na^+ channels selectively, thus depressing the rate of rise of the action potential (AP). The K^+ channels (Narahashi et al., 1964) or the slow Ca^{2+} and slow Na^+ channels (Rougier et al., 1969; Shigenobu & Sperelakis, 1972; Sperelakis, 1972a; Sperelakis & Shigenobu, 1972; Coraboeuf et al., 1976) are not affected. (b) The tetraethylammonium ion, TEA^+ (Armstrong & Binstock, 1965), and Ba^{2+} ion (Sperelakis & Lehmkuhl, 1966) block the resting and voltage-activated K^+ channels without blocking the fast Na^+ channels. Thus Ba^{2+} depolarizes by decreasing resting g_K, and Ba^{2+} and TEA^+ prolong the AP by depressing the rise in g_K that terminates the AP. (c) Veratridine slows Na^+ inactivation, i.e., the turnoff of the voltage-activated g_{Na}, and thereby prolongs the AP. The inactivation gate of the fast Na^+ channel is presumably slowed kinetically or made inoperative by veratridine. (d) Manganese ion (Mn^{2+}) and lanthanum ion (La^{3+}) block the slow Ca^{2+} channels preferentially. (e) Verapamil and D-600 (vasodilator and antiarrhythmic agents) block most slow channels (including slow Na^+ channels) without affecting the fast Na^+ channels (Kohlhardt et al., 1972; Shigenobu et al., 1974; Schneider & Sperelakis, 1975). (f) Zinc ion (Zn^{2+}), cadmium ion (Cd^{2+}), and hydrogen ion (H^+, with a low pH of about 6.1) block Cl^- conductance g_{Cl} in vertebrate skeletal muscle without affecting g_K.

(g) Low pH (about 6.0) also blocks slow Ca^{2+} channels (of cardiac muscle) without substantially affecting the fast Na^+ channels (Chesnais et al., 1975; Kohlhardt et al., 1976; Vogel & Sperelakis, 1977). (h) In resting membrane (frog skeletal muscle), temperature (Sperelakis, 1969) affects g_{Cl} less (Q_{10})

of 1.23) than g_K (Q_{10} of about 2.8). A low temperature depresses the membrane ion conductances, as it does for the conductivity of ionic solutions (the opposite of the situation for solid electrical resistors). (i) Local anesthetics (such as lidocaine, procaine, and cocaine) and diphenylhydantoin (anticonvulsant and antiarrhythmic agent) depress and block both fast Na^+ channels and slow channels (Josephson & Sperelakis, 1976). (Low concentrations, about 10^{-5} M of lidocaine and diphenylhydantoin, however, actually increase the inward fast Na^+ current but not the slow inward current.) The local anesthetics also decrease the resting K^+ conductance, and they depress the activity of the (Na-K)-ATPase.

The inner surface of the cell membrane is functionally different from the outer surface, as indicated by the following evidence: (a) Tetrodotoxin blocks fast Na^+ channels only when it is present in the outside bathing solution (Narahashi et al., 1964). (b) The addition of pronase, a proteolytic enzyme, to the solution perfusing the inside of a squid giant axon destroys the inactivation gate of the fast Na^+ channels, hence causing prolonged APs; thus the inactivation gate is located at the inner surface of the membrane (Armstrong & Binstock, 1965). (c) Tetraethylammonium ion acts to block K^+ channels at the inner surface of the membrane (Armstrong & Binstock, 1965). TEA^+ presumably fits inside the vestibule of the K^+ channel underneath the activation gate located at the inner surface of the membrane; i.e., the gate must be open before TEA^+ can get in to block. (d) Although the polarity of the resting potential in the perfused squid giant axon can be reversed by reversing the ionic gradients, an AP cannot be produced under these conditions. However, current can flow in reverse direction through an ion channel when the electrochemical driving force is reversed.

Chapter VIII

SURFACE CHARGE, STABLE STATES, AND OSMOTIC EFFECTS

A. Potential Profile Across Membrane

The cell membrane has fixed negative charges at its outer and inner surfaces. The charges are presumably due to acidic phospholipids in the bilayer and to protein molecules either embedded in the membrane (islands floating in the lipid bilayer matrix) or tightly adsorbed to the surface of the membrane. Most proteins have an acid isoelectric point, so that at a pH near 7.0 they possess a net negative charge. The charge at the outer surface of the cell membrane, with respect to the solution bathing the cell, is known as the zeta potential. This charge is responsible for the electrophoresis of cells in an electric field, the cells moving toward the anode (positive electrode), because unlike charges attract. (Some protozoa, e.g., paramecia, swim toward the cathode when they are alive, so-called galvanotaxic behavior, but are electrophoresed toward the anode when they are killed.) The surface charge affects the true potential difference (PD) across the membrane, as illustrated in Figure VIII-1A. At each surface, the fixed charge produces an electric field that extends a short distance into the solution and causes each surface of the membrane to be slightly more negative (by a few millivolts) than the extracellular and intracellular solutions. The potential theoretically recorded by an ideal electrode as the electrode is driven through the solution perpendicular to the membrane surface should become negative as the electrode approaches within a few Angstrom units of the surface; that is, the potential difference between the membrane surface and the solution declines exponentially as a function of distance from the surface. The length constant is a function of the ionic

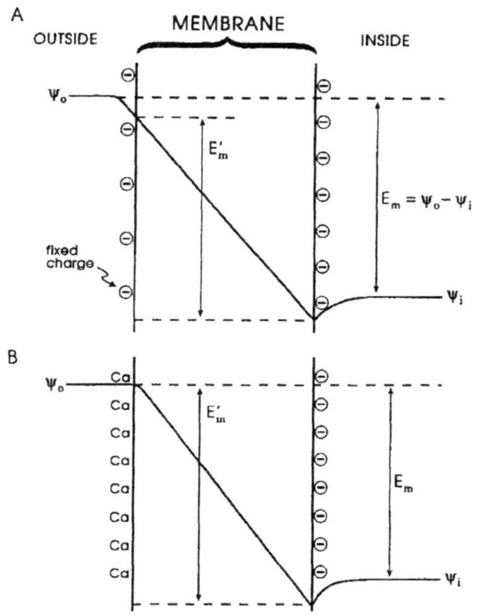

FIGURE VIII-1. Potential profile across the cell membrane. **A:** Because of fixed negative charges (at pH 7.4) at outer and inner surfaces of the membrane, there is a negative potential (indicated as downward) that extends from the edge of the membrane into the bathing solution on both sides of the membrane. Outer surface potential is known as the zeta potential and accounts for the passive electrophoresis of dead cells toward the anode. This surface potential falls off exponentially with distance into the solution, the length constant being determined by several factors, including ionic strength of the solution. Magnitude of the surface potential is a function of the charge density. ψ_o is the electrical potential of the outside solution, ψ_i is that of the inside solution, and membrane potential (E_m) is the difference (ψ_o - ψ_i). E_m is determined by the diffusion potentials and relative conductances. Profile of the potential through the membrane is shown as linear (the constant field assumption), although this need not be true for the present purpose. If the outer surface potential is exactly equal to that at the inner surface, then the true transmembrane potential (E_m') is exactly equal to the (microelectrode) measured membrane potential (E_m). **B:** If the outer surface potential is different from the inner potential, in this example done by elevating the extracellular Ca^{2+} concentration to bind Ca^{2+} to more of the negative charges, then the E_m' is greater than the measured E_m. Because the membrane ion channels are controlled by the E_m', this has an effect on function of the membrane (such as the threshold potential and inactivation potentials). Diminution of the inner surface charge decreases E_m'.

strength (or resistivity) of the solution: the lower the ionic strength, the greater the length constant is. The magnitude of the PD depends on the density of charge sites (number per unit of membrane area); the number of charges is also affected by the ionic strength and pH.

In Figure VIII-1A, the membrane potential measured by an intracellular microelectrode (E_m) is the potential of the outer solution (Ψ_o, the reference electrode) minus the potential of the inner solution (Ψ_i, the active microelectrode):

$$E_m = \Psi_o - \Psi_i$$

The true PD across the membrane ($E_m{}'$), however, is really that PD directly across the membrane, as shown in the figure. If the surface charges at each surface of the membrane are equal, then $E_m{}' = E_m$. If the outer surface charge is decreased to zero by extra binding of cations, such as Ca^{2+}, then the membrane becomes slightly hyperpolarized ($E_m{}' > E_m$), although this is not measurable by the intracellular microelectrode (Fig. VIII-1B). Conversely, if the outer surface charge were restored, and if the inner surface charge were neutralized, then the membrane would become slightly depolarized ($E_m{}' < E_m$); again this change is not measurable by the microelectrode. In Figure VIII-1 it has been assumed that the electric field through the thickness of the membrane is linear (the constant field assumption), although this need not be true for the present discussion.

Because the membrane ionic conductances are controlled by the PD directly across the membrane (i.e., by $E_m{}'$, and not by E_m) changes in the surface charges (for example, by drugs, ionic strength, or pH) can lead to apparent shifts in the threshold potential and in the E_m vs. h_∞ curve. For example, elevated $[Ca^{2+}]_o$ is known to raise the threshold potential (i.e., the critical depolarization required to reach electrical threshold), as expected from the small increase in E_m' that should occur. The apparent mechanical threshold (the mechanical threshold is the E_m value at which contraction of muscle just begins) can also be shifted by a similar mechanism. Finally, perfusing a squid giant axon internally with a sucrose solution of low ionic strength increases the internal surface charge and allows a larger PD than is measured by a microelectrode to exist across the membrane.

B. Multiple Stable States of Membrane Potential

The membrane potential of excitable cells, including myocardial cells, can possess more than one stable value. One stable condition, of course, is the normal resting potential for the cell. Cardiac Purkinje fibers, for example, exhibit two levels of resting potential, one at about -90 mV and the other at about -50 mV (Carmeliet, 1961; Hutter & Noble, 1961; Wiggins & Cranefield, 1976). Reconstruction of the Purkinje fiber action potential predicted two stable resting potentials, one at -88 mV and the second at -38 mV (McAllister et al., 1975). Switching between these two stable states (Wiggins & Cranefield, 1976) can occur in normal Ringer's solution but is facilitated in Na^+-free solutions containing tetraethylammonium ion (TEA^+) and tetramethylammonium ion (TMA^+). Potential state 1 (-90 mV) can be switched to potential state 2 (-50 mV) by brief depolarizing current pulses that are greater than a certain threshold level. Potential state 2 can be switched to state 1 by brief hyperpolarizing current pulses that are greater than a certain threshold. Sometimes spontaneous shifting between states occurs. The lower potential state (state 2) is a low g_K state (E_m farther from E_K), whereas the higher potential state (state 1) is a high g_K state (E_m closer to E_K). Measurements of membrane resistance in the two states confirms this. The transition from state 2 to state 1 (-50 mV to -90 mV) depends on a voltage-dependent increase in g_K. Transition from state 1 to state 2 (-90 mV to -50 mV) is caused by a sustained decrease in g_K in the presence of a background steady inward current carried by Na^+ and Ca^{2+}. Fibers in state 2 (-50 mV) are more readily excited and often possess automaticity sustained by depolarizing afterpotentials (cf. Sperelakis, 1972a).

In the study by Wiggins and Cranefield (1976), repetitive activity was seen in fibers that had a membrane potential in the vicinity of -50 mV only if g_K was low; that is, depolarizing a fiber from -90 to -50 mV by elevation of $[K]_o$ did not induce repetitive firing, because of the concomitant elevation of g_K in high $[K]_o$. This agrees with the findings on cultured heart cells (Sperelakis & Lehmkuhl, 1966). Upon cell separation the ventricular myocardial cells go into a low P_K state, which lowers the resting potential to about -52 mV and leads to automaticity.

In the low potential state (state 2), a small increase in $[K]_o$ produces a large hyperpolarization (Wiggins & Cranefield, 1976), presumably because of an increase in g_K triggered by the increase in $[K]_o$ (see section D in Chapter VI: *Resting Potential as Function of $[K]_o$*). Further elevation of $[K]_o$, of course, depolarizes, because of the concomitant decrease in E_K. A similar phenomenon has been demonstrated for cultured chick heart cells in a low P_K state with a low resting potential and a high R_m (Sperelakis & Lehmkuhl, 1966; Pappano & Sperelakis, 1969). An elevation of $[K]_o$ hyperpolarizes the cells. In some isolated cells in culture, P_K becomes so low that the cells are quiescent because the resting potential (about -15 mV) is too low to support action potentials (voltage inactivation of the slow cation channels); during the application of small hyperpolarizing current pulses, however, spontaneous action potentials with accompanying contractions occur (Pappano & Sperelakis, 1969).

Another observation related to the concept of multiple stable potential states is that Purkinje fibers and myocardial cells can exhibit small, slowly rising spikes superimposed on the plateau component of the action potential. In cultured heart cells (Sperelakis, 1972a) the addition of Ba^{2+} (0.5-5.0 mM) to lower g_K lengthens the plateau and gives rise to the superimposed spikes. (Ba^{2+} and K^+ have nearly the same unhydrated radii, 1.35 and 1.33 Å, respectively, and so Ba^{2+} presumably can fit into, but plugs, K^+ channels because the divalent charge of Ba^{2+} prevents it from readily passing through the K^+ channels.) Often this situation progresses to the point that a long-lasting train of small spikes rides on the low potential state. Termination of the train by a hyperpolarizing current pulse shifts the potential back to the high potential state. In smooth muscle, Ba^{2+} changes the shape of the action potential from the normal spike to a shape like that of a cardiac action potential with a prolonged plateau (Sperelakis & Tarr, 1965). With higher Ba^{2+} concentrations and longer times the membrane shifts into a sustained plateau state (of about -20 mV). Thus the plateau state may be considered as a third stable state.

Consistent with the idea that the plateau state is a stable state in the phenomenon of all-or-none repolarization in cardiac muscle, brief anodal (hyperpolarizing) current pulses applied during the middle of the plateau component can bring about sudden and premature repolarization to the resting potential; that is, the action potential can be terminated prematurely, provided that the hyperpolarizing pulse is suprathreshold. Thus bringing E_m artificially to some more negative level by a current pulse leads to sudden rapid repolarization (an increase in g_K). Hence the membrane potential shifts between two stable states: the plateau state and the resting state.

The cause of the spontaneous shifts from the high potential state 1 to low potential state 2 and back are not known. But a small Cl^- current, due to Cl^- passive redistribution according to the new potential, could produce slow changes in E_m that then trigger the other stable state when a threshold is reached. For example, if the potential is shifted from -90 to -50 mV by a brief depolarizing pulse, a net Cl^- influx occurs because of the new driving force (E_m–E_{Cl}); this Cl^- current is hyperpolarizing and thus could act as an applied hyperpolarizing current pulse to bring the membrane to a threshold point. In addition, because an increase in $[Ca^{2+}]_i$ causes an increase in g_K (Meech, 1974; Isenberg, 1975), this increase could contribute to a shift from the low potential state to the high potential state.

In the high potential state of -80 to -90 mV, the inward current during the fast-rising action potential is carried by Na^+ through the so-called fast Na^+ channels that are blocked by tetrodotoxin (TTX). In the low potential state of -50 mV, the fast Na^+ channels should be almost completely inactivated, and therefore the inward current of the slowly rising action potentials or slow responses is carried by Na^+ or Ca^{2+}, or both, through kinetically slow channels that are insensitive to TTX. The propagation velocities of these two types of action potentials are different, as expected. Because the low potential state tends to give rise to repetitive discharge, various cardiac arrhythmias can arise from situations in which there are two levels of resting potential. Indeed, a strong case has been built for cardiac arrhythmias arising in fibers in which a lowering of resting potential causes the fast-rising action potential to be replaced by a slowly rising response (Cranefield, 1975).

C. Osmotic Effects

The resting potential of myocardial cells is hyperpolarized by 5–10 mV in Ringer's solutions made twofold hypertonic by the addition of sucrose or extra NaCl (Hermsmeyer et al., 1972). In frog sartorious fibers also, the mean resting potential is slightly increased by a few millivolts, although the increase is not statistically significant (Sperelakis et al., 1970). Hypertonicity, however, reduces propagation velocity slightly in frog sartorious fibers and markedly in cardiac muscle. The relative velocity of propagation in a simple cylindrical cable of fixed length, whose internal resistivity varies in proportion to the volume change, should vary directly with the fourth root of the relative osmotic pressure of the bathing medium (Sperelakis et al., 1970) (see Appendix VIII-2). Because the observed depression was much greater in cardiac muscle than that predicted, the extra effect was attributed to an increased separation of the membranes at the intercalated discs. With sucrose hypertonicity, there was little or no depression of the maximum rate of rise of the cardiac action potential (but the peak overshoot potential was usually lowered). The duration of the action potential was generally markedly shortened. All of these effects were reversible.

The hyperpolarization produced in twofold hypertonic solution (e.g., in sucrose) probably results from the cell shrinkage. If each cell behaved as a perfect osmometer, which is almost true in hypertonic solutions (Page & Storm, 1965; Sperelakis & Rubio, 1971), then the cell volume would decrease to one-half. Thus the internal ion concentrations would double, whereas the external concentrations would remain the same. Hence E_K should increase by log 2 times 60 mV or by about 18 mV. This increase is more than enough to account for the observed hyperpolarization of 5–10 mV.

[Although frog sartorius fibers in hypertonic solution behave as though they have an osmotically inactive volume of about 33% (see Sperelakis & Rubio, 1971), i.e., they do not shrink in diameter to the extent predicted by ideal osmometer behavior, this behavior does not enter into this calculation because the myoplasmic water volume should be one-half.] A similar hyperpolarization of the nodal cells could account for the reported decrease in the spontaneous heart rate of isolated perfused hearts when the osmotic pressure is increased (Koch-Weser, 1963).

In agreement with these findings, Akiyama and Fozzard (1975) observed a mean hyperpolarization of only 4.6 mV for rabbit papillary muscle equilibrated in hypertonic solution (1.58 × isotonicity using sucrose); this hyperpolarization was much less than the predicted value of 10.6 mV, assuming that the muscle behaved as a perfect osmometer and that the activity coefficient of intracellular K^+ remained constant. With the assumption of perfect osmometer behavior, they calculated the ratio of the activity coefficient in hypertonic solution to that in isotonic solution ($\gamma_{1.58}/\gamma_{1.0}$) for a wide variety of $[K]_o$ values and obtained an average value of about 0.76. From this value, they concluded that the activity coefficient for intracellular K^+ declines substantially in hypertonic solution (as the internal K^+ concentration increases); thus the increase in internal K^+ activity ($\alpha_K{}^i$) in hypertonic solution is not as great as predicted and thereby accounts for the smaller than expected hyperpolarization (Table VIII-1).

Similar results were obtained by Fozzard and Lee (1976) with the same preparation (Table VIII-1). Their findings were also in agreement with those of Gibbons and Fozzard (1971), who found a mean hyperpolarization of 7.5 mV in sheep atrial trabeculae bathed in 1.7 x isotonic solution

TABLE VIII-1

Summary of data of resting membrane potential, internal K^+ activity, and internal K^+ activity coefficients obtained for rabbit papillary muscle in isotonic, hypotonic, and hypertonic solutions.

		Tonicity of Solution		
	Parameter Measured	**Hypotonic 0.76 x**	**Isotonic 1.0 x**	**Hypertonic 1.5 x or 1.58 x**
A	Internal K^+ activity coefficient ($\gamma^i{}_K$)	0.972 x 0.742	0.742	0.802 x 0.742
	Internal K^+ activity ($\alpha^i{}_K$)	0.738 x 119 mM	119 mM*	1.268 x 119 mM
	Resting potential (E_m)	-78.5 mV	-86.1 mV	-91.6 mV
B	Internal K^+ activity ($\alpha^i{}_K$) measured	60.5 mM	83.4 mM	110.4 mM
	Internal K^+ activity ($\alpha^i{}_K$) calculated #	63.6 mM	---	125 mM##
	Resting potential	-69.5 mV	-74.7 mV	---

Reference for part A is Akiyama & Fozzard, 1975, and that for part B is Fozzard & Lee, 1976.

In part A, the values entered are for a K_o of 3.1 mM.

*Calculated from intracellular K^+ concentration measured by extrapolation of the E_m vs. log $[K]_o$ curve to zero potential.

#Calculated assuming perfect osmometer behavior.

##Correction of K^+ activity coefficient for expected increase in internal ionic strength gives a value of 118 mM.

Gibbons and Fozzard (1971) found a mean hyperpolarization of 7.5 mV in sheep atrial trabeculae bathed in 1.7 x isotonic solution and ratio of 0.78 for the activity coefficient in the hypertonic solution to that in the isotonic solution. From the work of Hermsmeyer et al. (1972) on cat, guinea pig, chick, and frog ventricular muscles bathed in 2 x isotonic solution, Akiyama and Fozzard (1975) calculated a ratio of about 0.76 for the activity coefficient in the hypertonic solution to that in the isotonic solution.

and calculated a similar $\gamma_{1.7} / \gamma_{1.0}$ ratio of 0.78. Akiyama and Fozzard (1975) calculated a $\gamma_{2.0} / \gamma_{1.0}$ ratio of about 0.65 from the work by Hermsmeyer et al. (1972) on cat, guinea pig, chick, and frog ventricular muscles. By extrapolating the E_m vs. log $[K]_o$ curve to zero potential, Akiyama and Fozzard obtained a $[K]_i$ value of 160 mM (in isotonic solution), and multiplying this by the average γ value for internal K^+ (0.742), they obtained an internal K^+ activity ($\alpha_K{}^i$) value of 119 mM.

Hypotonic solutions, i.e., diluted Ringer's, (Sperelakis & Rubio, 1971) produce a pronounced swelling of myocardial cells that is actually greater than that predicted on the basis of an ideal osmometer. (Cell swelling or shrinking can be measured by electron microscopy or can be monitored simply by changes in muscle weight.) The resting potential diminishes markedly in 0.75 × isotonicity and becomes about -45 mV in 0.5 × isotonicity (Sperelakis et al., 1970). Propagation velocity, as expected from the depolarization, decreases markedly, going to zero by 0.5 × isotonicity. In frog sartorius fibers the depolarization and depression of propagation velocity produced by hypotonic solutions is considerably less. Thus in frog skeletal muscle the observed depolarization of about 14 mV in 0.5 × isotonicity can be entirely accounted for by the expected 18 mV decrease in E_K (60 mV times log 2, since cell volume doubles if the osmometer behavior is perfect). The larger depolarization in myocardial cells probably is related to the greater swelling produced.

In contrast to these results, Akiyama and Fozzard (1975) found that rabbit cardiac muscle (papillary) bathed in hypotonic (0.76 × isotonicity) solution depolarizes only 6.8 mV; this value was close to the predicted value of 6.9 mV that was based on the assumption that the muscle behaved like a perfect osmometer and that the activity coefficient of intracellular K^+ remained constant.

The tonicity of the bathing solution also affects the membrane ion conductances. For example, in frog sartorius fibers the hypertonicity decreases g_{Cl} and increases g_K whereas hypotonicity increases g_{Cl} but has only little effect on g_K (Sperelakis & Schneider, 1968). These results may be due to two effects: 1) an increase in the value of P_{Cl}, in the sarcolemma with an increase in fiber volume, and hence an increase in the stretch of the cell membrane; and 2) a marked swelling of the transverse (T) tubular system in hypertonic solution and a slight shrinking of the T tubules in hypotonic solutions. Assumptions made in this analysis were that the T-tubule membrane is impermeable to Cl^- and that the cell surface membrane is several times more permeable to Cl^- than to K^+ (Hodgkin & Horowicz, 1959; Adrian & Freygang, 1962; Sperelakis & Schneider, 1968). The results indicated that g_{Cl} varied linearly with the reciprocal of the relative tonicity and therefore with fiber volume. Thus, the Cl^- permeability of the surface membrane increases as the degree of stretch increases, whereas P_K did not appear to change. This behavior is consistent with the finding that the resting channels for Cl^- permeation have different properties than those for K^+, g_{Cl} having a low Q_{10} of about 1.23 (about that for diffusion of ions in free solution) and g_K having a high Q_{10} of about 2.8 (Sperelakis, 1969). Therefore, it is conceivable that stretch can increase the number of available Cl^- channels or their activity. The P_{Cl} also seems to be related to the cholesterol content of the membrane, a low cholesterol giving a low P_{Cl} (see Chapter I, section A).

D. K^+ Effects

It has been demonstrated that increasing the concentration of K^+ in the bathing medium to an optimal value leads to an increase in propagation velocity in cardiac muscle, but not in skeletal muscle which showed only a decline in velocity (Sperelakis et al., 1970). The increase in propagation velocity at the optimal $[K]_o$ level in cardiac muscle (but not skeletal muscle), cannot be accounted for by the small decrease in resting E_m, which should decrease the critical depolarization necessary to reach the threshold potential. If so, at least part of the effect may be due to an action on the intercalated disk membranes.

K^+ accumulation in the intercalated disk clefts may play some role in the transmission process. Depolarization of the prejunctional cell during the rising phase of its action potential should cause a rapid and large efflux of K^+ to occur at all surfaces because of the increased electrochemical driving force (assuming that the K^+ anomalous rectification does not occur instantaneously). Because of the restricted diffusion space at the disk, with limited access to the bulk extracellular fluid, the K^+ concentration in the junctional cleft, $[K]_{jc}$, may be rapidly elevated. A buildup of $[K]_{jc}$ would decrease E_K across the postjunctional membrane and could help to depolarize it to threshold (Sperelakis et al., 1985). This in turn, because of its large surface area, could help to bring the surface sarcolemma to threshold. $[K]_{jc}$ could then be rapidly lowered by diffusion of K^+ out of the mouth of the cleft and by K^+ pumping into the prejunctional and postjunctional cells; ATPase activity has been demonstrated at the intercalated disk membranes (De Beyer et al., 1962; Schulze and Wollenberger, 1969).

An increase in $[K]_{jc}$ to about 30 mM would be sufficient to produce a depolarization of 30 mV. It has been estimated

(Hoffman and Cranefield, 1960; Weidmann, 1956) that $[K]_o$ surrounding the entire myocardial cell may even increase to levels as high as 15 mM during the action potential plateau, and there is evidence that accumulation of K^+ outside the cell accounts for inhibition of automaticity following periods of increased frequency of stimulation (Alanis and Benitez, 1967; Lu et al., 1965). Other examples of K^+ accumulation in restricted diffusion spaces include those in the T tubules of muscle fibers during repetitive spiking (Freygang et al., 1964) and in the narrow clefts between neurons and Schwann cells (Frankenheuser and Hodgkin, 1956) or glial cells (Kuffler, 1967). Thus, anything which increases the width of the intercalated disk cleft, such as hypertonicity, could depress transmission because it would be more difficult to elevate $[K]_{jc}$ to a critical level; an increase in $[K]_o$ to an optimal level should allow the critical level to be reached earlier.

E. pH Effects

The pH has an effect on membrane ion conductances. For example, in frog skeletal muscle fibers the g_{Cl} is diminished in acid pH (Harris, 1965; Hutter & Warner, 1967; Sperelakis, 1969). This effect occurs in buffer systems that mainly vary extracellular pH (Hutter & Warner, 1967) and for a bicarbonate-CO_2 buffer that varies both extracellular and intracellular pH (Sperelakis, 1969). There was no effect on g_{Cl} between pH 7.4 (5% CO_2) and 7.1 (30% CO_2), but a pronounced decrease in g_{Cl} occurred at pH 6.4 (50% CO_2). The g_{Cl} went to zero at pH 6.1 (100% CO_2); i.e., R_m had the same value as in Cl^--free solution. Because pH had no effect on R_m in Cl^--free solution, it appears that there is little or no affect on g_K in acid pH (as low as 6.1 or 5.0). Hutter and Warner (1967) found that the relation between g_{Cl} and extracellular pH was sigmoidal between a pH of 5.0 and 9.8, with the steepest part of the curve occurring near pH 7.0, and they concluded that the charge on an imidazole grouping (histidine has a pK_a of 6.0) controlled the Cl^- permeability of frog sartorious membrane. They also pointed out that simple protonation of membrane pores at an acid pH should cause an increase, rather than the observed decrease, in g_{Cl}. [Increased positive charges on membrane pores, however, could account for the effect of acid pH on g_{Cl} in crustacean muscle, in which pH has an effect that is the opposite of that on frog sartorius muscle (Reuben et al., 1962).]

In cardiac muscle and Purkinje fibers, an acid pH also decreases g_K, as indicated by a reduction in anomalous rectification and in delayed rectification and by prolongation of the action potential (Chesnais et al., 1975; Coraboeuf et al., 1976). The natural g_{Cl} is normally low in heart (Carmeliet, 1961), and therefore a low pH cannot produce a large decrease in g_{Cl}. Alkaline pH increases g_K (E. Carmeliet and E. Coraboeuf, personal communications). Hence it appears that an acid pH decreases membrane permeability to both anions (Cl^-) and cations (K^+), whereas alkaline pH increases the permeability to both. This action further indicates that a simple change in the charge of membrane pores cannot account for the changes in permeabilities.

In cardiac muscle it has been demonstrated that acid pH has a pronounced depressant effect on the inward slow Ca^{2+} current. For example, in embryonic chick ventricular muscle, 50% inhibition of the inward slow current occurs at about pH 6.5, and the slow current is completely abolished at about pH 6.0 (Vogel & Sperelakis, 1977). Because the rate of rise of the normal action potential, which is dependent on an inward fast Na^+ current, is only slightly affected at these same pH levels (although electromechanical uncoupling occurs as expected from the inhibition of the slow channels), acid pH quite selectively blocks slow cation channels rather than fast Na^+

channels. In frog atrial fibers studied by voltage clamp, Chesnais et al. (1975) found that the slow Ca^{2+} current was reduced by only 40% at a pH of 5.6 but that the slow Na^{+} current was completely abolished. Similar results were obtained by Kohlhardt et al. (1976) using cat papillary muscle, in which they demonstrated that the slow inward current decreased to less than half at pH 5.5 and that the slow channels recovered more slowly from inactivation.

The hydrogen ion also has a pronounced effect on the membrane (Na-K)-ATPase. For example, the (Na-K)-ATPase from chick embryo heart is 50% inhibited at a pH of about 6.3, and the inhibition is about 75% at pH 5.5 (Sperelakis & Lee, 1971).

Appendix VIII

Theoretical Velocity as a Function of Tonicity in a Cable

In hypertonic solution, a marked decrease in propagation velocity occurs in cardiac muscle, whereas only a slight decrease occurs in skeletal muscle (Sperelakis et al., 1970). Since propagation velocity in cardiac muscles decreased drastically in 2× hypertonic solution, even in those muscles presumably nearly devoid of T tubules and gap junctions, it seems unlikely that alterations in the properties of these structures mediate the decrease in velocity. In addition, there were no substantial changes in resting potential or max dV/dt to account for the large decrease in velocity. The changes in the cable properties in 2× hypertonic solution cannot account for the large depression of action potential (AP) velocity in cardiac muscle, but may account for the small depression in skeletal muscle (see below). Therefore, hypertonicity may have a major effect on the functioning of the intercalated disks. Hypertonicity has been shown to increase the average cleft width of the intercalated disks of cardiac muscle (Barr et al., 1965; De Mello et al., 1969; Dreifuss et al., 1966; Sperelakis and Rubio, 1971).

The observed velocity data for frog sartorius fibers bathed in 2× sucrose hypertonic solutions fit nearly exactly the predicted values based on the experimentally-measured R_i values, which actually increase rather than decrease (see Table VIII-A-1). If an increase in R_i in hypertonic solutions were more marked in cardiac muscle than in skeletal muscle, then this could be a factor involved in the greater decrease in propagation velocity in cardiac muscles compared to frog sartorius.

TABLE VIII-A-1. Summary of theoretical calculations giving velocity of propagation of cat cardiac muscle and frog sartorius at various tonicities.

Parameter	Volume Changes	Frog Sartorius	Cat Papillary
I. Volume Ratios			
A. Hypertonic			
$\frac{Vol_{2\times}}{Vol_{1\times}}$	Theoretical Measured	$1/2.0 = 0.50$ 0.66^{*}	$1/2.0 = 0.50$ $0.41^{\#}$
$\frac{Vol_{2.7\times}}{Vol_{1\times}}$	Theoretical Measured		$1/2.7 = 0.37$ $0.41^{\Diamond}$
B. Hypotonic			
$\frac{Vol_{0.75\times}}{Vol_{1\times}}$	Theoretical Measured		$1/0.75 = 1.33$ $1.74^{\#}$
$\frac{Vol_{0.84\times}}{Vol_{1\times}}$	Theoretical Measured		$1/0.84 = 1.19$ $1.43^{\Diamond}$
$\frac{Vol_{0.5\times}}{Vol_{1\times}}$	Theoretical Measured	$1/0.50 = 2.00$ 1.63^{*}	
II. Theoretical Velocity Ratios			
A. ΔRadius and ΔR_i *(eq 6)*			
$\frac{\theta_{2\times}}{\theta_{1\times}} = \sqrt[4]{\frac{Vol_{1\times}}{Vol_{2\times}}}$	Theoretical Measured	$\sqrt[4]{1/0.50} = 1.19$ $\sqrt[4]{1/0.66} = 1.11$	$\sqrt[4]{1/0.50} = 1.19$ $\sqrt[4]{1/0.41} = 1.25$
$\frac{\theta_{0.75\times}}{\theta_{1\times}} = \sqrt[4]{\frac{Vol_{1\times}}{Vol_{0.75\times}}}$	Theoretical Measured		$\sqrt[4]{1/1.33} = 0.93$ $\sqrt[4]{1/1.74} = 0.87$

Table VIII-A-1 cont.

Parameter	Volume Changes	Frog Sartorius	Cat Papillary
$\frac{\theta_{0.5\times}}{\theta_{1\times}} = \sqrt[4]{\frac{Vol_{1\times}}{Vol_{0.5\times}}}$	Theoretical Measured	$\sqrt[4]{1/2.00} = 0.84$ $\sqrt[4]{1/1.63} = 0.89$	
B. ΔRadius and $\Delta R_i'$ (eq. 10A and 10B).			
1. For hypertonic (eq. 10B)			
$\frac{\theta_{2\times}}{\theta_{1\times}} = \sqrt[4]{\frac{Vol_{1\times}}{Vol_{2\times}}} \cdot \sqrt{0.85}$	Theoretical Measured	$0.92\sqrt[4]{1/0.50} = 1.09$ $0.92\sqrt[4]{1/0.66} = 1.02$	$0.92\sqrt[4]{1/0.50} = 1.09$ $0.92\sqrt[4]{1/0.41} = 1.15$
2. For hypotonic (eq. 10A)			
$\frac{\theta_{0.75\times}}{\theta_{1\times}} = \sqrt[4]{\frac{Vol_{1\times}}{Vol_{0.75\times}}} \cdot \sqrt{\frac{1}{0.85}}$	Theoretical Measured		$1.09\sqrt[4]{1/1.33} = 0.98$ $1.09\sqrt[4]{1/1.74} = 0.95$
$\frac{\theta_{0.5\times}}{\theta_{1\times}} = \sqrt[4]{\frac{Vol_{1\times}}{Vol_{0.5\times}}} \cdot \sqrt{\frac{1}{0.85}}$	Theoretical Measured	$1.09\sqrt[4]{1/2.00} = 0.91$ $1.09\sqrt[4]{1/1.63} = 0.82$	
C. ΔRadius only, R_i unchanged (eq. 11)			
$\frac{\theta_{2\times}}{\theta_{1\times}} = \sqrt[4]{\frac{Vol_{2\times}}{Vol_{1\times}}}$	Theoretical Measured	$\sqrt[4]{0.50} = 0.84$ $\sqrt[4]{0.66} = 0.90$	$\sqrt[4]{0.50} = 0.84$ $\sqrt[4]{0.41} = 0.90$
$\frac{\theta_{0.75\times}}{\theta_{1\times}} = \sqrt[4]{\frac{Vol_{0.75\times}}{Vol_{1\times}}}$	Theoretical Measured		$\sqrt[4]{1.33} = 1.07$ $\sqrt[4]{1.74} = 1.15$
$\frac{\theta_{0.5\times}}{\theta_{1\times}} = \sqrt[4]{\frac{Vol_{0.5\times}}{Vol_{1\times}}}$	Theoretical Measured	$\sqrt[4]{2.00} = 1.19$ $\sqrt[4]{1.63} = 1.13$	

Table VIII-A-1 cont.

Parameter	Volume Changes	Frog Sartorius	Cat Papillary
D. ΔRadius, meas R_i inc 16% in 2× hyper (eq. 12)			
$\frac{\theta_{2x}}{\theta_{1x}} = \sqrt[4]{\frac{\mathrm{Vol}_{2x}}{\mathrm{Vol}_{1x}}} \cdot \sqrt{\frac{1}{1.16}}$	Theoretical Measured	$0.93\sqrt[4]{0.50} = 0.78$ $0.93\sqrt[4]{0.66} = 0.84$	
III. Measured velocity ratios			
$\frac{\theta_{2x}}{\theta_{1x}}$		0.86	0.62 0.20 (Chick) 0.51 (Frog)
$\frac{\theta_{0.75x}}{\theta_{1x}}$			0.34 0 (Chick)
$\frac{\theta_{0.5x}}{\theta_{1x}}$		0.72	0 0.55 (Frog)

* Values taken from Sperelakis and Schneider, 1968. ♦Values taken from Sperelakis et al., 1970. ⊛Values taken from Page and Storm, 1965. ✧R_i' is the theoretical R_i value corrected for the fact that the relationship between conductivity of ionic solutions and concentration on a log-$\log_{10}$ plot has a slope less than 1.0. For Ringer solution, the slope was measured to be 0.80 for a 10-fold increase and a 10-fold decrease in ion concentration (Sperelakis, 1969). The corresponding value for NaCl between 10 and 100 mM is 0.90 from the published tables in the *Handbook of Chemistry and Physics* (Hodgman et al., 1958). Therefore, a correction factor of 0.85 was used for the above calculations.

Most of the depolarization of the cardiac and skeletal muscles in hypotonic solution can be accounted for by the gain of water and swelling of the cells, thus lowering $[K]_i$, hence E_K and the resting E_m. If the cells behaved as perfect osmometers, cell volume would increase to 200% in 0.5× solution and decrease to 50% in 2× solution. Thus, dilution of $[K]_i$ to one-half can produce a maximum depolarization of about 18 mV (decrease in E_K of log 2 × 61 mV), which is sufficient to account for the depolarization of frog sartorius fibers. A hyperpolarization of the theoretical 18 mV was not observed in either the cardiac or skeletal muscles bathed in 2× hypertonic solutions (Sperelakis and Schneider, 1968; Sperelakis et al., 1970), probably because either there is an increased deviation from a purely K^+-selective membrane at the larger resting potentials and/or there is some loss of intracellular K^+. The latter is consistent with the observed lack of decrease in R_i of frog sartorius fibers (Sperelakis and Schneider, 1968).

Approximately 0% osmotically inactive volume has been measured in cat papillary muscle bathed in hypertonic solutions (Page and Storm, 1965; Sperelakis and Rubio, 1971). In contrast, frog sartorius fibers obey the changes predicted from the Van't Hoff relation only after correction for an osmotically inactive volume of about 33% (Sperelakis and Schneider, 1968). Thus, the ratio of diameters in 2×/1× tonicities is 0.81, and the ratio of volumes is $(0.81)^2$ or 0.66, and not the theoretical 0.50. The corresponding theoretical calculations for each category corrected for the measured volume changes are given in Table VIII-A-1.

The following demonstrates that the relative velocity of propagation in a simple cylindrical cable of fixed length, and whose internal resistivity varies in proportion to the volume change, should vary directly with the 4th root of the relative osmotic pressure of the bathing medium. Velocity of propagation of the action potential (θ) in a simple cable may be obtained by the following analysis (Hodgkin, 1964; Hodgkin and Huxley, 1952b; Katz, 1948; Katz, 1966). For a single fiber bathed in a large volume, the membrane current density during propagation of an AP is:

$$I_m = \frac{a}{2}\frac{1}{R_i}\frac{\partial^2 V}{\partial x^2} \qquad \text{eq. A-1}$$

where I_m = membrane current density (amp/cm^2), a = radius of cell (cm), R_i = specific resistance of sarcoplasm (ohm-cm), and x = distance along the fiber.

If the AP propagates at a constant velocity (θ), then

$$x = \theta \cdot t \qquad \text{eq. A-2}$$

Substituting into eq. 1 gives:

$$I_m = \frac{a}{2}\frac{1}{R_i}\frac{1}{\theta^2}\frac{d^2 V}{dt^2} \qquad \text{eq. A-3}$$

Therefore

$$\theta^2 = \frac{a}{2}\frac{1}{R_i}\frac{1}{I_m}\frac{d^2 V}{dt^2} \qquad \text{eq. A-3a}$$

and

$$\theta = \sqrt{\frac{a}{2}\frac{1}{R_i}\frac{1}{I_m}\frac{d^2V}{dt^2}} \qquad \text{eq. A-4}$$

Since

$$I_m = C_m \frac{dV}{dt} + I_i \qquad \text{eq. A-5}$$

Therefore, conduction velocity is also an inverse function of C_m , which is implicit in the I_m term in *equation 1*. For the following analysis, it is assumed that C_m remains constant, the specific membrane properties remain the same for fibers at different tonicities, and folding of the surface membrane does not occur (cf. Martin, 1954). If so, then I_m and d^2V/dt^2 remain constant (Katz, 1966). Hence,

$$\theta \propto \sqrt{\frac{a}{R_i}} \qquad \text{eq. A-2}$$

Therefore,

$$\frac{\theta_{n\times}}{\theta_{1\times}} = \sqrt{\frac{a_{n\times}}{a_{1\times}}}\sqrt{\frac{R_{i1\times}}{R_{in\times}}} \qquad \text{eq. A-3}$$

where the subscript n× refers to one value of relative tonicity and 1× to isotonicity. Upon increase in osmotic pressure of the bathing solution (Π), the cells shrink. For cylindrical cells, diameter should vary as the square root of cell volume (Vol), assuming that length does not change. Changes in R_i should vary directly with changes in cell volume because of dilution or concentration of the ions in the myoplasm. Therefore, for cases in which there is perfect osmometer behavior and R_i changes in proportion to ΔVol (r_i remains constant):

$$\frac{\theta_{n\times}}{\theta_{1\times}} = \sqrt[4]{\frac{\mathrm{Vol}_{n\times}}{\mathrm{Vol}_{1\times}}}\sqrt{\frac{R_{i1\times}}{R_{in\times}}} \qquad \text{eq. A-4}$$

$$= \sqrt[4]{\frac{\mathrm{Vol}_{n\times}}{\mathrm{Vol}_{1\times}}}\sqrt{\frac{a^2{}_{1\times}}{a^2{}_{n\times}}}$$

By substituting into *equation* 4,

$$\frac{\theta_{n\times}}{\theta_{1\times}} = \sqrt[4]{\frac{\mathrm{Vol}_{n\times}}{\mathrm{Vol}_{1\times}}}\sqrt{\frac{\mathrm{Vol}_{1\times}}{\mathrm{Vol}_{n\times}}} \qquad \text{eq. A-5}$$

$$= \left(\frac{\mathrm{Vol}_{1\times}}{\mathrm{Vol}_{n\times}}\right)^{-0.25}\left(\frac{\mathrm{Vol}_{1\times}}{\mathrm{Vol}_{n\times}}\right)^{+0.50}$$

$$= \left(\frac{\mathrm{Vol}_{1\times}}{\mathrm{Vol}_{n\times}}\right)^{+0.25}$$

Therefore,

$$\frac{\theta_{n\times}}{\theta_{1\times}} = \sqrt[4]{\frac{\mathrm{Vol}_{1\times}}{\mathrm{Vol}_{n\times}}} = \sqrt[4]{\frac{\Pi_{n\times}}{\Pi_{1\times}}} \qquad \text{eq. A-6}$$

Thus, if the muscle cells behaved as perfect osmometers, then velocity should vary directly with the 4th root of the relative osmotic pressure or increase by 19% in 2× hypertonicity ($\sqrt[4]{2}$ = 1.19) (see Table VIII-A-1).

The passive velocity of propagation in a cable (θ_p) is constant when $t/\tau_m >> 1$, and is given by (Cole, 1968):

$$\theta_p = \frac{2}{\sqrt{r_m}\, c_m \sqrt{r_i}} = \frac{1.414\sqrt{a}}{\sqrt{R_m}\, C_m \sqrt{R_i}} \qquad \text{eq. A-7}$$

Thus, if the sarcolemma were to fold in hypertonic solution, such that the surface area per unit fiber length remained constant, then r_m and c_m , as well as r_i , would remain unchanged; hence passive velocity would be independent of the tonicity. For example, conduction velocity of frog sartorius muscle was found to be nearly independent of muscle length, and this fact was accounted for on the basis of a folded-membrane hypothesis (Martin, 1954). Velocity of the action potential is similar to the passive velocity (Hodgkin, 1964; Cole, 1968).

The conductivity of NaCl and KCl solutions, however, increases only about 9.0-fold for a 10-fold increase in concentration (from 10 to 100 mM) (Hodgman et al., 1958). The measured resistivity of Ringer solution (ρ_{1x}) diluted 10-fold actually increases only 8.0-fold, and that of Ringer solution concentrated 10-fold decreases only 8.0-fold (Sperelakis, 1969). Thus, if an average correction factor of 0.85 is used, the corresponding relationships become:

$$\frac{\rho_{n\times}}{\rho_{1\times}} = \frac{R_{in\times}}{R_{i1\times}} = k\left(\frac{\mathrm{Vol}_{n\times}}{\mathrm{Vol}_{1\times}}\right) \qquad \text{eq. A-8}$$

where $k = 1/0.85$ if $n\times < 1\times$,

and $k = 0.85$ if $n\times > 1\times$

For cases in which R_i changes in proportion to ΔVol, but corrected for the relationship between conductivity and concentration (R_i'):

$$\frac{\theta_{n\times}}{\theta_{1\times}} = \sqrt[4]{\frac{\mathrm{Vol}_{n\times}}{\mathrm{Vol}_{1\times}}}\sqrt{\frac{R_{i1\times}\,k}{R_{in\times}}} \qquad \text{eq. A-9}$$

$$= \sqrt{k}\sqrt{\frac{\Pi_{n\times}}{\Pi_{1\times}}}$$

Since $\sqrt{1/0.85} = 1.085$ and $\sqrt{0.85} = 0.922$:

$$\frac{\theta_{n\times<1\times}}{\theta_{1\times}} = 1.085\sqrt[4]{\frac{\Pi_{n\times}}{\Pi_{1\times}}} \qquad \text{eq. A-10A}$$

$$\frac{\theta_{n\times>1\times}}{\theta_{1\times}} = 0.922\sqrt[4]{\frac{\Pi_{n\times}}{\Pi_{1\times}}} \qquad \text{eq. A-10B}$$

If R_i remains unchanged but a changes as if the cell were a perfect osmometer (from equation 4):

$$\frac{\theta_{n\times}}{\theta_{1\times}} = \sqrt[4]{\frac{\mathrm{Vol}_{n\times}}{\mathrm{Vol}_{1\times}}} = \sqrt[4]{\frac{\Pi_{1\times}}{\Pi_{n\times}}} \qquad \text{eq. A-11}$$

However, instead of a decrease in R_i in proportion to the experimentally-measured volume decrease in 2× sucrose hypertonic solution, it was found (Sperelakis and Schneider, 1968) that R_i of frog sartorius fibers actually increased about 16% in Cl^--containing solution. (In 0.5× hypotonic solutions, R_i increased about 58%, which is about as expected (Sperelakis and Schneider, 1968).) Thus, for special case of 2× sucrose hypertonic solution in which an actual measured increase in R_i (R_i'') of 16% occurs rather than the theoretical decrease in R_i:

$$\frac{\theta_{2\times}}{\theta_{1\times}} = \sqrt[4]{\frac{\Pi_{1\times}}{\Pi_{2\times}}}\sqrt{\frac{R_{i1\times}}{R_{i2\times}}} \qquad \text{eq. A-12}$$

$$= \sqrt[4]{\frac{\Pi_{1\times}}{\Pi_{2\times}}}\sqrt{\frac{1}{1.16}}$$

$$= 0.928\sqrt[4]{\frac{\Pi_{1\times}}{\Pi_{2\times}}}$$

Thus, if R_i remained unchanged (equation A-11) or were to actually increase slightly in 2× hypertonic solution (equation A-12), then velocity should vary inversely with the 4th root of the relative tonicity, or decrease to 84% in 2× hypertonicity ($\sqrt[4]{1/2} = 0.84$).

Chapter IX

ELECTROGENIC SODIUM PUMP POTENTIALS

As described in several preceding chapters, the Na-K pump is responsible for maintaining the cation electrochemical gradients. The diffusion potentials for K^+ (E_K) and Na^+ (E_{Na}) are about -94 mV and +60 mV, respectively (although some estimates of E_{Na} go as high as +130 mV). The resting potential value is usually near E_K, because the K^+ permeability (P_K) is much greater than P_{Na} in a resting membrane. The exact resting membrane potential (E_m) depends on the P_{Na}/P_K ratio, some types of cells having P_{Na}/P_K ratios of .01–.05, whereas that of others (e.g., smooth muscle or nodal cells of the heart) have a ratio closer to 0.15. In most muscle and nerve cells, the resting E_m has a smaller magnitude (i.e., is less negative) than E_K by 5–30 mV. Therefore at first glance it would appear as though there were no electrogenic pump potential contribution to the resting potential; that is, as though the Na-K pump could be only indirectly responsible for the resting potential by its role in producing the ionic gradients.

A direct contribution of the pump to the resting E_m, however, can often be demonstrated. For example, if the Na-K pump is blocked by the addition of ouabain (which has no significant effect on the conductances), there usually is an immediate depolarization of 1–35 mV, depending on the cell and on the conditions. For most vertebrate myocardial and skeletal muscle cells, the immediate depolarization upon inhibition of the pump is only about 2–8 mV; that is, the contribution of an electrogenic pump to the measured resting E_m is small under physiological conditions. But under conditions in which the pump is stimulated to pump at a high rate — i.e., under conditions in which $[Na]_i$ is abnormally high — the direct electrogenic contribution of the pump to the resting potential is usually much greater (e.g., 10–30 mV or even more), and E_m can actually exceed E_K by as much as 30 mV or more.

A. Na^+ Loading for Pump Stimulation

In virtually all muscle and nerve cells tested — both vertebrate and invertebrate — if the ionic electrochemical gradients are allowed to run down (e.g., by storing the tissues at low temperatures for several hours), then after allowing the tissues to restart pumping, the measured E_m usually exceeds the calculated E_K by 10–30 mV for a period of time (e.g., Kernan, 1962; Cross et al., 1965; Frumento, 1965; Kernan, 1970); see Figure IX-1. The Na^+ loading of the muscle cells can be facilitated by placing them in cold low-K^+ or zero-K^+ Ringer's solutions, external K^+ being necessary for the Na-K linked pump to operate; K_m of the (Na-K)-ATPase for K^+ is about 2 mM. After several hours in such a solution the internal concentrations of Na^+, K^+, and Cl^- approach the concentrations in the bathing Ringer's solution, and the resting potential is low (< -40 mV). The tissue is then transferred to a pumping solution, which is the appropriate Ringer's solution containing normal K^+ and at normal temperature (e.g., -25° C for tissues from cold-blooded animals and 37° C for those from warm-blooded animals).

Under such conditions, the pump turns over at nearly a maximal rate, because the major control over pump rate is $[Na]_i$. (In some cell types the low initial E_m also might stimulate the pump rate, because the energy required to pump out Na^+ is much less.) The measured E_m of such Na^+ preloaded cells increases rapidly, more rapidly than E_K (proportional to log $[K]_i/[K]_o$), as shown in Figure IX-1. After this transient phase, however, a crossover of the two curves occurs, so that E_K again exceeds E_m , as in the physiological condition. Any

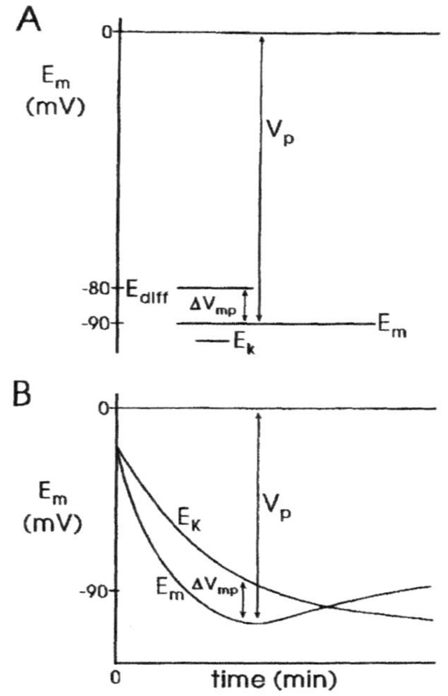

FIGURE IX-1. Diagrammatic representation of an electrogenic sodium pump potential. **Upper graph:** Muscle cell whose net ionic diffusion potential (E_{diff}, function of diffusion potentials and relative conductances) is -80 mV but yet exhibits a measured membrane resting potential (E_m) that is greater (-90 mV in illustration). Difference between E_m and E_{diff} (10 mV in example) represents the contribution of the electrogenic pump to the resting potential. Usual direct contribution of the pump is only a few millivolts and can be measured by the amount of depolarization produced immediately after complete inhibition of the (Na-K)-ATPase by cardiac glycosides. The question, however, is whether the electrogenic pump potential (V_p) should be considered to be 10 mV (V_p') or 90 mV (V_p''). Because the pump pathway is separate from and parallel to the ionic conductance pathways, the electrogenic pump potential must be 90 mV for this example. **Lower graph:** Cell that was run down (Na loaded, K depleted) over several hours by inhibition of Na-K pumping, resulting in a low resting potential. Returning the muscle cell to a pumping solution allows the resting E_m to build back up as a function of time. Buildup in E_m occurs faster than buildup in E_K as illustrated. Whenever E_m is greater (more negative) than K^+ potential (E_K), the difference (V_p') must reflect (as a minimum) the contribution of the sodium pump potential. Again, the actual pump potential may be represented by V_p''. Buildup in E_m may not occur instantaneously with repumping, because of a low membrane resistance (R_m) in the run-down muscle cell.

method used to increase membrane resistance (R_m) enhances the electrogenicity of the pump, i.e., causes a greater difference between E_m and E_K during the transient phase (e.g., see Adrian & Slayman, 1966; Harris & Ochs, 1966). The reader is referred to an article by Luttgau and Glitsch (1976) for a complete summary of all the types of cells for which this kind of experiment has been done and for the findings.

For example, Page and Storm (1965) found that rewarming cat papillary muscles previously cooled to 2° C led to the rapid restoration of the normal resting potential (within 10 min), whereas recovery of the intracellular Na^+ and K^+ concentrations was much slower. Glitsch (1972) found the same phenomenon in guinea pig atrial cells and demonstrated that cardiac glycoside reversed the transient hyperpolarization beyond E_K; there was a correlation between the active Na^+ efflux and the degree of hyperpolarization (Glitsch, 1972). In a corollary type of experiment, McDonald and MacLeod (1971) observed that during a prolonged period of hypoxia the resting potential of guinea pig papillary muscles decreases much less than E_K (a difference of about 25 mV). In such a situation, it appears that an electrogenic pump acts in an attempt to hold the resting potential constant despite dissipating ionic gradients; that is, even though the pump rate is slowing, the degree of electrogenicity of the pump (Na-K coupling ratio) might be increasing.

1. Na^+ Injection

Another method often used to demonstrate that the pump can be electrogenic is to inject Na^+ ions into the cell through the voltage-recording microelectrode. This method was first reported by Hodgkin and Keynes in 1956 for squid giant axon; they observed a small hyperpolarization and concluded that an increased active Na^+ efflux might contribute directly to the membrane potential. Nishi and Soeda (1964) injected Na^+

into toad spinal ganglion cells and observed a transient large hyperpolarization. Kerkut and Thomas (1965) injected Na^+ into snail neurons and observed a rapid hyperpolarization (up to 30 mV) that was immediately abolished by ouabain, and they showed that changes in membrane resistance, which could account for the hyperpolarization, did not occur. Using a voltage clamp technique to measure the current produced by the electrogenic Na^+ pump in snail neurons, Thomas (1969) found that the pump current and the rate of Na^+ extrusion rose in proportion to the amount of Na^+ injected. The reader is again referred to the article by Luttgau and Glitsch (1976) for a summary of this type of experiment.

2. Criteria for Electrogenicity

It is often difficult to prove that the pump is electrogenic. First, it must be unequivocally demonstrated that, for example, hyperpolarization produced upon Na^+ injection is not the result of enhanced pumping of an electroneutral pump. Such a neutral pump, in some situations, could cause depletion of external K^+ in a restricted diffusion space just outside the cell membrane (e.g., a periaxonal space), the depletion thus leading to a larger E_K and thereby to hyperpolarization. In such a situation, depletion occurs because the Na-K pump, linked 3 : 3, pumps in K^+ faster than replenishment (by diffusion from the bulk interstitial fluid) occurs. Such K^+ depletion could occur, for example, in small nonmyelinated nerve fibers surrounded by Schwann cells.

In such a situation, when the pump is blocked by ouabain or by metabolic poisons, one could not distinguish between electrogenic and nonelectrogenic pumping. Second, if hyperpolarization is really due to an electrogenic Na^+ pump, the hyperpolarization must be augmented under conditions that increase membrane resistance. In fact, the normally high membrane resistance is probably one major reason for the large effect of the stimulated Na^+ pump on membrane potential in molluscan neurons. Third, it must be demonstrated that ionic conductance changes (e.g., an increase in g_K or a decrease in g_{Na}) cannot account for the observed hyperpolarization. This possibility could be ruled out whenever E_m exceeds (is more negative than) E_K. It is often difficult to accurately determine E_K, however, because it depends on the accurate measurement of $[K]_i$, a measurement that often poses problems. For example, the free K^+ concentration, and actually the K^+ activity, must be known. The activity coefficient for internal K^+ appears to be somewhat lower than that for external K^+ (Lee & Fozzard, 1975) and could conceivably change in different situations.

3. Effect of Membrane Potential

Several reports have suggested that the electrogenic Na^+ pump may be influenced by the membrane potential (Fozzard & Kipnis, 1967; Horowicz & Gerber, 1965). From energetic considerations, depolarization should enhance the electrogenic Na^+ pump, whereas hyperpolarization should inhibit. Because depolarization reduces the electrochemical gradient (and hence the energy requirements) against which Na^+ must be extruded, whereas hyperpolarization increases the gradient, this behavior can be deduced. If the energetics are important, there should be a distinct potential, more negative than E_K, at which Na^+ pumping is prevented. Such a pump equilibrium potential was found to be about -95 mV in Na^+-loaded frog sartorius fibers upon rewarming (Tahara et al., 1973); on the other hand, some investigators have observed a high pumping rate in Na^+-loaded frog muscle when E_m is about -95 mV (Adrian & Slayman, 1966; Sjodin & Beauge, 1968). The pump equilibrium potential was 20 mV (or more) more negative than the resting potential in snail neurons (Kostyuk et al., 1972); in some snail neurons, however, the pump current was less voltage dependent and seemed to be generated by a constant current source (Kostyuk et al., 1972). In addition, others (e.g., Thomas, 1969; Marmor, 1971) have reported that membrane potential does not have strong direct effects on the pump rate in molluscan neurons.

4. Effect of Cl^-

There seems to be no evidence that the net pump current is directly affected by Cl^-, i.e., that Cl^- ions accompany the pumped Na^+ or pumped K^+ in close proximity (e.g., on the carrier itself) and thereby alter the net current. For example, if 3 Na^+ plus 1 Cl^- were extruded for every 2 K^+ pumped in, then the net current would be zero. In general, however, Cl^- ions are known to have a short-circuiting effect on the electrogenic Na^+ pump potential. For example, if the external Cl^- is replaced by less permeant anions (such as sulphate or isethionate), the magnitude of the post-tetanic hyperpolarization produced in mammalian nonmyelinated nerve fibers, demonstrated to be brought about by the activation of an electrogenic Na^+ pump mechanism during tetanic stimulation, is greatly increased (Rang & Ritchie, 1968). The mechanism of the Cl^- short-circuiting (i.e., lowering the magnitude of the pump potential) is not known, but the short-circuiting could be caused by the general lowering of membrane resistance in the presence of Cl^- ; as stated in subsection 2 on *Criteria for Electrogenicity*, the greater R_m is, the greater the electrogenic pump potential is. An alternate way to view this is that the Cl^- distribution acts as a clamp trying to hold E_m at E_{Cl}.

B. Density, Turnover Number, and Pump Current

The density of Na-K pump sites, estimated by specific binding of [^{3}H]ouabain, in a wide variety of cells is usually about 1,000/μm^2. For example, Landowne and Ritchie (1970) estimated a value of about 750/μm^2 for mammalian nerve. The turnover rate of the pump is generally estimated to be 20-100 sec^{-1}. For example, a value of 22 sec^{-1} (at 20° C) was obtained in mammalian nerve (Landowne & Ritchie, 1970). The pump current (I_p) has been measured from Ohm's law ($I_p = V_p R_m$), where V_p is the pump potential, and from voltage clamp experiments. Estimates of the pump current are about 10 pmol/cm^2 per sec (e.g., see Christoffersen, 1972, for snail neuron). A density of 1,000 sites/μm^2 (10^{11} sites/cm^2) times a turnover rate of 20/sec gives 2×10^{12} turnovers/cm^2 per sec. If 3 Na^+ are pumped with each turnover, this gives 6×10^{12} Na^+ ions/cm^2 per sec; dividing by Avogadro's number (6.02×10^{23} ions/mol) yields 10×10^{-12} mol/cm^2 per sec, which is the same number as the 10 pmol/cm^2 per sec measured. Of course, the net pump current is less, depending on the amount of K^+ pumped in the opposite direction (i.e., on the coupling ratio).

C. Model and Equivalent Circuit

Because 3 Na^+ are known to be transported per molecule of ATP spent, we can assume that the pump has 3 negatively charged binding sites for Na^+ at the inner surface of the membrane. We also can assume that the pump has 3 negatively charged binding sites for Na^+ at the inner surface of the membrane. We also can assume that the pump has a much higher affinity for Na^+ than for K^+ at the inner surface sites, whereas the converse is assumed to be true for the negatively-charged binding sites at the outer surface. We assume that the pump cannot operate unless fully loaded with 3 Na^+ ions. If the pump is not electrogenic, i.e., no net charge is moved across the membrane, then 3 K^+ ions must be carried in for every 3 Na^+ ions moved out (coupling ratio of 3 : 3). In this model, the pump cannot complete the cycle unless fully loaded with 2 or 3 K^+ ions. If the pump is to be electrogenic, i.e., produce a net current (and hence potential) across the membrane, then the amount of K^+ pumped in must be less than the amount of Na^+ pumped out; e.g., the Na-K coupling ratios must be 3 : 2 or 3 : 1. The coupling ratio cannot be 3 : 0, because of the well-known fact that external K^+ must be present for the pump to operate. In a wide variety of cells under conditions in which the pump is electrogenic, the most usual coupling ratio reported is 3 : 2. The coupling ratio apparently can be altered, however, in the same cell under different conditions. Pump carrier models have been proposed (Cross et al., 1965).

The pump site actually consists of one giant protein complex that completely spans the membrane. The pump consists of two subunits: alpha and beta (Fig. IX-2A). The alpha subunit is larger (ca. 112 kDa) and protrudes predominantly on the cytoplasmic side of the membrane, whereas the beta subunit (ca 35 kDa) protrudes mainly on the outer surface and contains several glycosylation sites. The molecular structures proposed for the alpha and beta subunits are illustrated in Figure IX-2B. As depicted, the alpha subunit has seven membrane-spanning domains with alpha-helix structure. These domains are hooked together by connecting links that are alternately intracellular and extracellular. Several of the intracellular loops are very long, especially those between domains 4 and 5 and domains 2 and 3. The amino terminus is intracellular, and the carboxy terminus is extracellular, as shown. The beta subunit has only one membrane-spanning domain, with a very long extracellular loop containing the carboxy terminus and several glycosylation sites. It is not known whether the $\alpha\beta$-unit (monomeric) or an ($\alpha\beta)_2$-unit (dimeric) is the physiological organization in the cell membrane (Cornelius, 1991).

1. Coupling Ratio

Whenever the Na-K pump is stimulated to turn over faster, e.g., by increasing $[Na]_i$ or $[K]_o$, for a given

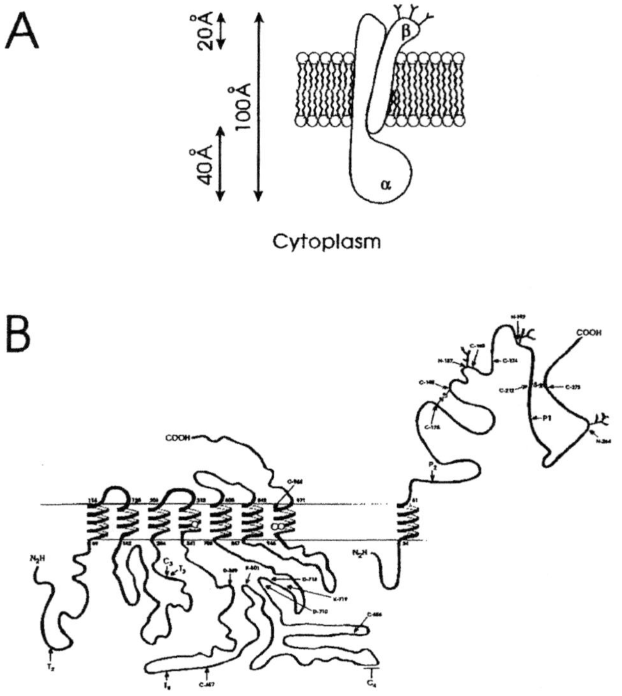

FIGURE IX-2. Models for the Na-K-ATPase/Na-K pump of the cell membrane. **A:** Model depicting the arrangement of the alpha and beta subunits (α, β protomers). As shown, both sides of the alpha subunit and the extracellular side of the beta subunit protrude well beyond the borders of the lipid bilayer, and the extracellular portion of the beta subunit contains glycosylation sites. **B:** Molecular structure of the Na-K-ATPase based on hydropathy analysis of the alpha and beta subunits. As depicted, the alpha subunit (on the left) has seven membrane-spanning domains containing alpha-helix structure, and with extracellular/intracellular connecting links. Some of the intracellular links consist of long loops (bottom). The amino terminus (intracellular) is at the left, and the carboxy terminus (extracellular) is at the top right. There are over 971 amino acids in the alpha subunit. The beta subunit has only one membrane-spanning domain, with a very long extracellular loop containing the carboxy terminus and several glycosylation sites. There are 302 amino acids in the beta subunit. The amino terminus is intracellular. [Figures reproduced from Pedemonte and Kaplan (1990) with permission, and were originally taken from Kirley (1989).]

coupling ratio (e.g., 3 : 2), the electrogenic pump potential contribution to E_m becomes larger, as described in section A on *Na^+ Loading for Pump Stimulation.* The pump potential contribution also becomes larger, for a constant pumping rate, if the coupling ratio is increased (e.g., to 3 : 1). It is not certain which factors control the coupling ratio. For example, it is not even known whether the coupling can be 3 : 3 or whether it must always be some fixed ratio such as 3 : 2 (see Thomas, 1972). For squid giant axon, Mullins and Brinley (1969) concluded that the Na^+-K^+ coupling ratio was close to 1.0 at a low $[Na]_i$ and that the ratio increased to about 1.5 and 3.0 as $[Na]_i$ was increased. Sjodin and Beauge (1968) suggested a ratio of 2.0 for squid axon, and Mullins and Noda (1963) gave a ratio of 3.0 for frog skeletal muscle. Others have reported that the ratio does change for frog skeletal muscle (Adrian & Slayman, 1966; Keynes & Rybova, 1963).

Sjodin and Ortiz (1975) found a coupling ratio of about 1.5 for frog sartorius muscle, but the ratio was changed to 1.1 in the presence of Ba^{2+} ions for reduction of K^+, conductance; they suggested that the membrane compensates for a reduced passive K^+ influx by producing a greater current of inwardly pumped K^+ ions. The net effective electrogenic pump current was reduced about sixfold with Ba^{2+}, about the same degree as the reduction in g_K , the similarity indicating that the K^+ ion is the main ion acting as a shunt on the pump potential. The ouabain-sensitive membrane potential was about the same in the presence or absence of Ba^{2+}, presumably because of the lower coupling ratio but higher membrane resistance.

Thomas (1969) reported that the coupling ratio (about 3 : 2) was independent of $[Na]_i$ for snail neurons. The Na-K pump in several cell types can switch from a Na^+-K^+ exchanging mode of operation to a Na^+/Na^+ exchanging mode of operation (e.g., see Baker et al., 1969a), and the mode is governed by the intracellular ATP : ADP : P_i ratios. In addition, there is some evidence that the Na-K coupling ratio may be affected by the ADP concentration (DeWeer, 1970).

2. Equivalent Circuit

It is difficult to draw an equivalent circuit for the situation in which an electrogenic Na^+ pump potential is present. Most discussions of electrogenic pump potentials mainly consider the contribution of the pump potential to the measured E_m , i.e., the difference in E_m when the pump is operating vs. immediately after the pump has been stopped by the addition of ouabain. Consequently it appears as though the pump potential (V_p) were in series with the net cationic diffusion potential (E_{diff}). (E_{diff} is the E_m that would exist solely on the basis of the ionic gradients and relative permeabilities in the absence of an electrogenic pump potential, as calculated from the constant field equation.) Indeed, the usual equation given (e.g., see Droogmans & Casteels, 1976) is

$$E_m = E_{\text{diff}} + R_m i_p$$

where i_p is the electrogenic component of the pump current. This equation states that E_m is the sum of E_{diff} and a voltage drop produced by the electrogenic pump.

The pump potential should be considered to be in parallel with E_{diff}. Because the density of pump sites is about $1{,}000/\mu m^2$, whereas that of Na^+ and K^+ channels in resting membrane is estimated to be more than 1,000-fold lower, there should be no relation between the pump pathway (the active flux path) and R_m (the passive flux paths). For example, high-K^+ red blood cells have a large density of Na-K pump sites, but a low permeability to Na^+ and K^+. Most membrane models propose that the pump path and the passive conductance paths are in parallel. A second important point is that the true pump potential must be much greater than the ΔE_m measured in the absence and presence of ouabain. Namely, the pump potential should really be considered as the full potential between zero and the maximum negative pump potential (V_p'') while the pump is pumping, as indicated in Figure IX-1. If the full pump potential were only a few millivolts, the pump could not transport Na^+ against an electrochemical gradient of 140–210 mV (E_m–E_{Na}).

One possible equivalent circuit for an electrogenic Na^+ pump, which takes into account many of the known facts, is given in Figure IX-3. First, the pump pathway is in parallel with the resistance pathways, and the pump potential is the full potential and not the ΔV. Second, the pump pathway is shown as essentially an infinite resistance (capacitance dielectric), so that it cannot act as a short-circuit path to E_{diff} when the pump potential is low or zero. Third, a relatively fixed resistance of moderate value in series with the pump (R_p) gives the circuit the feature that the observed pump potential contribution to E_m (ΔV) is a function of membrane resistance (R_m): the higher R_m is (R_p constant), the more nearly E_m approaches V_p. Fourth, the pump capacitor is charged to some voltage ($V_p = Q_p/C_p$) and is held there by a pump current generator (Na-K coupling ratio of 3 : 2, 3 : 1.5, or 3 : 1) in parallel with the capacitor. (The pump capacitor essentially acts as a variable-voltage battery, as long as the pump is operating.) Fifth, if the pump is stopped by ouabain, the pump potential rapidly dissipates through R_m with a time constant of $(R_p + R_m)\, C_m$. Sixth, if $V_p < E_{diff}$ (-80 mV), then E_m = -80 mV; i.e., the pump potential is hidden by E_{diff}. If $V_p > E_{diff}$, then E_m is somewhere between E_{diff} and V_p. The exact E_m measured depends on the magnitude of V_p and on R_m. The V_p increases with an increase in the rate of pumping and with an increase in the coupling ratio.

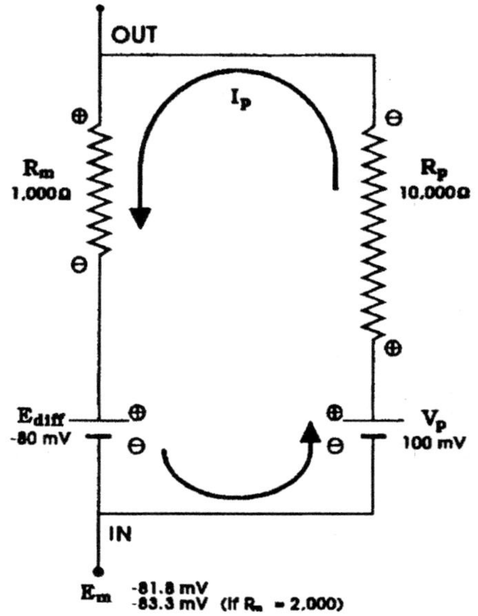

FIGURE IX-3. Hypothetical electrical equivalent circuit for electrogenic sodium pump. Model consists of a pump pathway in parallel with the membrane resistance (R_m) pathway (and membrane capacitance, C_m). This model fits the evidence that the pump is independent of short-range membrane excitability and that the pump proteins and channel proteins are embedded in the lipid bilayer as parallel elements. Net diffusion potential (E_{diff} determined by the diffusion potentials and relative conductances) of -80 mV is depicted in series with R_m. Pump leg is assumed to consist of a pump voltage generator (V_p) of 100 mV in series with a pump resistance, R_p. It is assumed that R_p does not change with changes in R_m, and that R_p is about 10 times R_m. Net electrogenic pump current (I_p) would be developed by the pumping in of only 2 K^+ ions for every 3 Na^+ ions pumped out. For the values given in the figure (namely, R_m of 1,000 Ω , E_{diff} of -80 mV, R_p, of 10,000 Ω, and V_p of -100 mV), by circuit analysis it may be calculated that the measured membrane potential (E_m) is -81.8 mV; that is, the direct electrogenic pump potential contribution to the resting potential is -1.8 mV. If R_m were raised to 2,000 Ω (e.g., by placing the membrane in low Cl^- solution, or by adding Ba^{2+} ion to decrease P_K), then the calculated E_m would be -83.3 mV, i.e., closer to V_p (-100 mV).

Using circuit analysis (see Appendix I), for the values of the parameters given in Figure IX-3, E_m would be -81.8 mV, i. e., moderately close to the value of V_p (-90 mV); see Table IX-1, Calculation Trial D. If R_m is now raised fivefold (to 5,000 Ω), E_m would be -86.7 mV, i.e., closer to V_p (Table IX-1, Calculation Trial E). Thus, this circuit clearly gives a pump potential contribution to E_m that is dependent on R_m. Another way to state this is that the higher R_m is (relative to R_p), the more E_m reflects V_p because of less short-circuiting of V_p by R_m. Finally, if E_{diff} is made smaller (e.g., in cells with a higher P_{Na}/P_K ratio under normal conditions or if E_{diff} were lowered by the elevation of $[K]_o$), then the relative contribution of the pump potential to E_m becomes greater. For example, for the same other parameter values indicated in Figure IX-3, if E_{diff} is changed from -80 mV to -50 mV, then E_m would be -66.7 mV; i.e., the ΔV is now much larger: 16.7 mV (Table IX-1, Calculation Trial F) vs. 1.8 mV (Table IX-1, Calculation Trial D). These and other combinations are summarized in Table IX-1. As expected, an increase of V_p (by stimulating the pump rate or by increasing the coupling ratio) hyperpolarizes (compare Calculation Trials D and C in Table IX-1), and the amount of hyperpolarization produced by an increase in V_p is greater when E_{diff} is smaller (compare Calculation Trials G and D in Table IX-1).

D. Physiological Importance

The physiological importance of an electrogenic pump potential is not fully known, particularly for myocardial cells. In most vertebrate striated muscle cells, the electrogenic pump potential contribution to the membrane resting potential is small (about 2–8 mV). Although small, however, this contribution could have significant effects on the level of inactivation of the fast Na^+ channels (see Chapter XII on *Effect of Resting Potential on Rate of Rise of Action Potential*).

TABLE IX-1

Summary of calculations of resting potential for one possible electrical equivalent circuit for an electrogenic pump potential in parallel with the net diffusion potential.

Calculation Trial	E_{diff} mV	R_m ohm-cm^2	V_p mV	E_m mV	ΔV mV
A	-80	1,000	0	-72.7	+7.3
B	-80	1,000	-80	-80.0	0
C	-80	1,000	-90	-80.9	-0.9
D	-80	1,000	-100	-81.8	-1.8
E	-80	5,000	-100	-86.7	-6.7
F	-50	5,000	-100	-66.7	-16.7
G	-50	1,000	-100	-54.5	-4.5

E_{diff} = Net diffusion potential. V_p = Electrogenic pump potential. $\Delta V = (E_m - E_{diff})$. The resistance in series with the pump (R_p) was held constant at 10,000 Ω-cm^2. V_p can be made zero by blockade of the pump with ouabain. V_p can be increased (e.g., -90 to -100 mV) by increasing the pumping rate or by increasing the coupling ratio. ΔV can be increased by an increase in V_p or in R_m. R_m can be increased by decreasing the Cl^- conductance of the membrane (e.g., with Cl^--free solution, by lowering pH to 6.1, or by addition of compounds that block Cl^- channels) or by decreasing K^+ conductance (e.g., by Ba^{2+}).

Further, for all excitable cells, one can imagine that an electrogenic pump potential would act to delay depolarization under adverse conditions (e.g., ischemia and hypoxia) and would act to speed repolarization of the normal resting potential during recovery. It is crucial that the excitable cell maintain its normal resting potential as much as possible, because of the effect on the AP's rate of rise and conduction velocity with small depolarizations, and because of the complete loss of excitability with larger depolarizations. Several investigators have concluded that an electrogenic Na^+ pump contributes to the membrane potential of heart tissues under physiological conditions (Hiraoka & Hecht, 1973; Isenberg & Trautwein, 1974; Wiggins & Cranefield, 1976). For example, in voltage clamp experiments on sheep Purkinje fibers, Isenberg and Trautwein (1974) found a time-independent outward constant current that was immediately inhibited by ouabain.

In cells in which there are lower resting potentials (e.g., smooth muscle cells and cardiac nodal cells) and in which the electrogenic pump potential can be considerably larger as demonstrated in subsection C2 on *Equivalent Circuit*, sinusoidal oscillations in the Na-K pumping rate could produce oscillations in E_m. Such oscillations could exert important control over the spontaneous firing of the cell; that is, the period of enhanced pumping hyperpolarizes the cell and suppresses automaticity, whereas slowing of the pump leads to depolarization and consequently to triggering of action potentials. Such a mechanism has been proposed for the slow large sinusoidal oscillations of smooth muscle (Prosser et al., 1976); these oscillations are suppressed by ouabain and enhanced by injection of Na^+. [In contrast, on the basis of data from rabbit intestinal muscle, it has been suggested that, under steady-state conditions, the electrogenic Na-K pump contributes only a few millivolts to the resting E_m and therefore is unlikely to account for the large oscillations in E_m (El-Sharkawy & Daniel, 1975).]

Oscillation of the pump may be brought about by some simple mechanism. For example, the firing of several spikes should raise $[Na]_i$ (smooth muscle cells have a small volume : surface area ratio), and stimulate the electrogenic pump; the increased pumping rate, in turn, hyperpolarizes and suppresses firing, thus allowing $[Na]_i$ to become lower again and removing the stimulation of the pump; the latter depolarizes and triggers spikes; and the cycle could be repeated. For such oscillations, all that would be required is a brief lag between elevation in $[Na]_i$ and the maximum pump rate. Noma and Irisawa (1974) concluded that rabbit sinoatrial nodal cells possess an electrogenic Na^+ pump and that the pump might be one factor that modulates the heart rate under physiological conditions. When stimulated at a high rate, cardiac Purkinje fibers and nodal cells undergo a transient period of inhibition of automaticity after cessation of the stimulation, a period known as **overdrive suppression**. Vassalle (1970) concluded that stimulation of an electrogenic pump due to elevation in $[Na]_i$ is the major cause of this phenomenon.

Finally, in small-diameter myelinated and nonmyelinated nerve fibers a brief period of repetitive stimulation leads to a long-lasting hyperpolarization known as **posttetanic (or postactivation) hyperpolarization**. The mechanism proposed for this phenomenon is the activation or stimulation of an electrogenic pump due to the elevation in $[Na]_i$ (Ritchie & Straub, 1957; Connelly, 1959; Rang & Ritchie, 1968). The resultant hyperpolarization has profound effects on the release of neurotransmitter at the nerve terminal to a subsequent impulse; if severe, the hyperpolarization could even produce conduction blockage along the axon. A similar phenomenon in receptor neurons could have pronounced physiological effects.

E. Measurement of Electrogenic Na-K Pump Current

In each reaction cycle, the Na-K pump transports $3Na^+$ out of and $2K^+$ into the cell for each ATP hydrolyzed (Eisner & Lederer, 1980; Rakowski et al., 1989), thus generating a net outward current (Gadsby et al., 1985). This electrogenic Na-K pump current (I_p) has been shown to vary with voltage and to decline with hyperpolarization (Nakao & Gadsby, 1986; Goldshlegger et al., 1987; Schweigert et al., 1988; Rakowski & Paxson, 1988; Bahinski et al., 1988). In whole-cell voltage clamped single ventricular cardiomyocytes of guinea pig, the reversal potential ($E_{Na/K}$) for I_p was about -150 mV and the current density was about 3 $\mu A/cm^2$ (at 0 mV) (Gadsby & Nakao, 1989).

In contrast to an ionic current, I_p is generated by a pump utilizing energy from ATP hydrolysis. Consequently, $E_{Na/K}$ depends not only on the electrochemical gradients for Na^+ and K^+, but also on the ATP level of the cell. $E_{Na/K}$ may be defined as the voltage at which the free energy of ATP hydrolysis is equal and opposite to the work required for the transfer of Na^+ and K^+ (De Weer et al., 1988). Assuming a stoichiometry of 3 Na^+ : 2 K^+:

$$E_{Na/K} = E_{ATP} + 3E_{Na} - 2E_K \qquad \text{eq. 1}$$

where E_{ATP} is the ATP driving potential defined as:

$$E_{ATP} = (E^0_{ATP} + RT/F) \ln \frac{[ADP][P_i]}{[ATP]} \qquad \text{eq. 2}$$

where E^0_{ATP} is the ATP driving potential under standard conditions. Hyperpolarization increases the electrochemical gradient for extrusion of Na^+, thereby increasing the energy cost. Hyperpolarization, therefore, should decrease I_p. Metabolic inhibition causing the intracellular ATP level to decrease reduces the energy source available for sodium pumping, and thereby shifts the voltage at which I_p vanishes ($E_{Na/K}$) to less negative potentials.

The (Na-K)-ATPase was found to exist in embryonic chick hearts as early as the third day of development, and its specific activity increases during development (Sperelakis & Henn, 1970; Sperelakis & Lee, 1971; Sperelakis, 1972b). An electrogenic pump potential contribution of a few millivolts to the resting potential has been shown in embryonic chick heart preparations (Pelleg et al., 1980), and the existence of an I_p at early stages of development was shown in young (3-day-old) embryonic chick heart cells using the whole-cell voltage-clamp technique (Meszaros, preliminary unpublished measurements).

I_p should be isolated by blocking all other ionic currents and Ca/Na exchange. This can be done by putting 50 mM NaCl and 20 mM TEA-Cl in the pipette solution, and 0 Ca^{2+}, 0 K^+, 2.0 mM $BaCl_2$, 1.0 mM CsCl, 1.0 mM $CoCl_2$, and 10 μM TTX. (to block the fast Na^+ current) in the external solution. The K^+ currents are blocked by the Ba^{2+}, Cs^+ and TEA, and the Ca^{2+} current is blocked by Co^{2+} and Ca^{2+}-free solution. Under these conditions, since all time-dependent conductances are blocked, the current/voltage relation remaining gives the **background current (I_{bg})** (Fig. IX-4A). The outward I_p can be activated by changing the K^+-free superfusate to 5.4 mM K^+-containing external solution, and blocked by ouabain (10 μM). The ouabain-sensitive I_p current is then obtained by subtracting

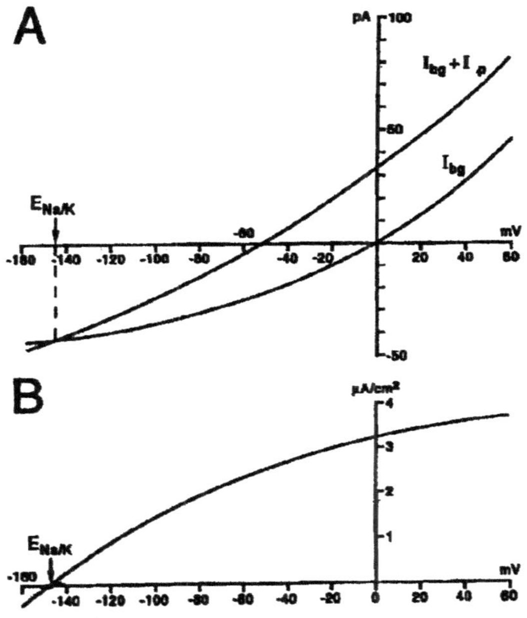

FIGURE IX-4. Illustration of technique for measuring electrogenic pump current (I_p) using whole-cell voltage clamp. Hypothetical curves are shown. **A:** Current/voltage relationships in a cardiomyocyte plotted using the current values measured at the end of applied long-duration pulses. The two curves give the background current (I_{bg}) recorded in 0 K^+ or in ouabain and the pump current (I_p) plus I_{bg} . $E_{Na/K}$ indicates the zero-current potential, which corresponds to the reversal potential for I_p. **B:** I_p current density ($\mu A/cm^2$) as a function of membrane potential in heart cells. Curve plotted is based on the difference current obtained by subtracting the two curves in part A.

currents recorded in the absence and presence of ouabain (in normal K^+ solution) (Fig. IX-4B). It was demonstrated that half-maximal activation of I_p is produced by about 10 mM Na^+ in the pipette, and over 90% activation by 50 mM (Nakao & Gadsby, 1989).

I_p declines to zero at about -144 mV, which is the zero-current potential for I_p. The density of I_p (at 0 mV) is about half in young myocytes compared to that in old myocytes, suggesting that the pump current density markedly increases during embryonic development. This is consistent with the reported increase in both (Na-K)-ATPase activity and the number of ouabain binding sites.

Using the above protocol for directly measuring I_p, the effect of various agents on the magnitude of I_p can then be assessed. For example, I_p can be measured in single cultured rat skeletal myotubes that are rounded up using colchicine treatment to destroy the microtubules. Using the same protocol as outlined above, one can measure I_p at different voltages. The zero-current potential was about -140 mV, and the current density (at 0 mV) was about 0.3 pA/pF or 0.3 $\mu A/cm^2$ (assuming a C_m of 1.0 $\mu F/cm^2$) (Li & Sperelakis, 1994). Thus, the effect of several agents that are believed to stimulate or inhibit the Na-K pump can be assessed. For example, insulin produces a substantial hyperpolarization of rat skeletal muscle in vivo and in vitro (Iannacone et al., 1989; Li and Sperelakis, 1993), presumably by stimulation of the electrogenic Na-K pump by means of the incorporation of spare membrane into the cell membrane. Beta-adrenergic agonists (e.g., isoproterenol) hyperpolarize skeletal muscle by stimulation of electrogenic Na-K pumping rate by phosphorylation of the existing (Na-K)-ATPases (Li and Sperelakis, 1993) (for other references, see Sperelakis, 1985).

Appendix IX

Electrogenic Pump Potential

If the number of Na^+ ions extruded by the Na-K pump exceeds the number of K^+ ions pumped in, then the Goldman constant field equation requires modification to allow for the current generated by the electrogenic pump. A term was inserted into the constant field equation by Mullins and Noda (1963) to take into account the operation of an electrogenic pump under steady-state conditions

$$E_m = \frac{-RT}{F} \ln \frac{bP_K[K]_i + P_{Na}[Na]_i}{bP_K[K]_o + P_{Na}[Na]_o} \qquad \text{eq. A-1}$$

where b is the coupling ratio of the Na-K pump. A derivation of this equation has also been given by Thomas (1972) and by Sjodin and Ortiz (1975). Using this equation, calculations indicate that the direct contribution of the electrogenic pump potential to the observed resting potential is greater (a) when the coupling ratio is higher, and (b) when the P_{Na}/P_K ratio is higher, resulting in a lower resting potential due to the ion electrochemical gradients (E_{diff}). Note that Equation A-1 never gives an E_m that is greater than E_K (see Table IX-A-1).

Sjodin and Ortiz (1975) derived an equation that applies during nonsteady-state conditions, such as those that apply during recovery of muscle from the Na^+-loaded state and that give E_m values more negative than E_K

$$E_m = \frac{RT}{F} \ln \frac{b\left(\frac{f_{Na}}{f_K}\right)[K]_o + P_{Na}[Na]_o}{b\left(\frac{f_{Na}}{f_K}\right)[K]_i + P_{Na}[Na]_i} \qquad \text{eq. A-2}$$

TABLE IX-A-1

Calculated resting potentials (in mV) based on the Mullins-Noda modification of the constant-field equation for different Na/K pump coupling ratios.

P_{Na}/P_K Ratio	Coupling Ratios				Differences in E_m		
	3 : 3	3 : 2	3 : 1.5	3 : 1	3 : 2	3 : 1.5	3 : 1
0.01	-86	-89	-90	-91	3	4	5
0.10	-54	-62	-67	-73	8	13	19
Difference	32	27	23	18			

For these calculations, the following values were used: 150 mM for $[K]_i$, 1 mM for $[Na]_i$, 4 mM for $[K]_o$, and 150 mM for $[Na]_o$. Cl^- was assumed to be passively distributed, and Ca^{2+} contribution to the resting potential was neglected. A coupling ratio of 1.0 (3 : 3) is, of course, the same as an absence of an electrogenic pump potential. The calculated E_K is -94.6 mV. E_m never becomes greater than E_K, even with a high coupling ratio of 3 : 1, by using this equation. [Equation from Mullins and Noda, 1963.]

or

$$E_m = \frac{RT}{F} \ln \frac{[\mathrm{K}]_o + \frac{1}{b} \frac{f_\mathrm{K}}{f_\mathrm{Na}} \frac{P_\mathrm{Na}}{P_\mathrm{K}} [\mathrm{Na}]_o}{[\mathrm{K}]_i + \frac{1}{b} \frac{f_\mathrm{K}}{f_\mathrm{Na}} \frac{P_\mathrm{Na}}{P_\mathrm{K}} [\mathrm{Na}]_i} \qquad \text{eq. A-2a}$$

where b is the coupling ratio of the Na-K pump, and f is the ratio of passive net flux, ϕ, to pumped flux, M ($f = \phi/M$). In the steady-state condition, $f_{Na} = f_K = 1$, and Equation A-2 reduces to the Mullins-Noda equation (Equation A-1). When $b = 1$, Equation A-1 reduces to the Goldman equation.

El-Sharkawy and Daniel (1975) also modified the Goldman constant field equation to derive an equation, similar to that of Sjodin and Ortiz, that holds under steady-state or non steady-state conditions, providing that the net membrane current is zero

$$E_m = \frac{RT}{F} \ln \frac{P_\mathrm{K} [\mathrm{K}]_o + y P_\mathrm{Na} [\mathrm{Na}]_o + P_\mathrm{Cl} [\mathrm{Cl}]_i}{P_\mathrm{K} [\mathrm{K}]_i + y P_\mathrm{Na} [\mathrm{Na}]_i + P_\mathrm{Cl} [\mathrm{Cl}]_o} \qquad \text{eq. A-3}$$

where $y = [1 - \gamma + (\gamma/b)]$, γ = active Na^+ efflux/passive Na^+ influx (M_{Na}/ϕ_{Na}), and b is the coupling ratio of the Na-K pump. Equation A-3 does not presume a passive distribution of Cl^- ions. When $\gamma = 0$ (no pumping) or when $b = 1$ (electroneutral pump), the calculated potential is equal to the Goldman potential.

Moreton (1969) used a different modification of the constant field equation to take into account the presence of an electrogenic pump

$$E_m = \frac{+RT}{F}\ln\frac{P_K[K]_o + P_{Na}[Na]_o + J_{ep}\frac{RT}{FE_m}}{P_K[K]_i + P_{Na}[Na]_i + J_{ep}\frac{RT}{FE_m}} \qquad \text{eq. A-4}$$

where J_{ep} is the net efflux of cations via an electrogenic Na-K pump. Equation A-4 was derived from the following considerations, which must apply during steady state [the reader is also referred to the derivation by Christoffersen (1972)]

$$J_{Na}^{p} = -J_{Na}^{a} \qquad \text{eq. A-5}$$

and

$$-J_{K}^{p} = J_{K}^{a} \qquad \text{eq. A-6}$$

where $-J_{Na}{}^{a}$ is the active Na^+ efflux, $J_{Na}{}^{p}$ is the passive Na^+ influx, $-J_K{}^{p}$ is the passive K^+ efflux, and $J_K{}^{a}$ is the active K^+ influx. (The sign convention used is positive when the electrical field assists the ion movement and negative when the electrical field opposes.) The coupling ratio (b) of the Na-K pump then is

$$b = \frac{-J_{Na}^{a}}{J_{K}^{a}} = \frac{J_{Na}^{p}}{-J_{K}^{p}} \qquad \text{eq. A-7}$$

Rearrangement gives

$$b\,J_{K}^{p} + J_{Na}^{p} = 0 \qquad \text{eq. A-8}$$

Mullins and Noda inserted equations for the passive net fluxes in Equation A-8 and obtained their Equation A-1. Subtraction of Equation A-6 from Equation A-5 gives

$$J^p_{\mathrm{Na}} + J^p_{\mathrm{K}} = -J^a_{\mathrm{Na}} - J^a_{\mathrm{K}} \qquad \text{eq. A-9}$$

or

$$J^p_{\mathrm{Na}} + J^p_{\mathrm{K}} = -\left(J^a_{\mathrm{Na}} + J^a_{\mathrm{K}}\right) = -J_{\mathrm{ep}} \qquad \text{eq. A-9a}$$

which means that the algebraic sum of the passive fluxes must equal the sum of the active fluxes, and the sum of the active fluxes must equal the net pump flux. The J_{ep} is negative when b is more than 1.0. The following two relationships for the passive net fluxes from the constant field equation

$$J^p_{\mathrm{Na}} = P_{\mathrm{K}} E_m \frac{F}{RT} \frac{[\mathrm{K}]_i e^{E_m F/RT} - [\mathrm{K}]_o}{1 - e^{E_m F/RT}} \qquad \text{eq. A-10}$$

and

$$J^p_{\mathrm{Na}} = P_{\mathrm{Na}} E_m \frac{F}{RT} \frac{[\mathrm{Na}]_i e^{E_m F/RT} - [\mathrm{Na}]_o}{1 - e^{E_m F/RT}} \qquad \text{eq. A-11}$$

were inserted into Equation A-9a to yield Equation A-4.

In Equation A-4, because P_{Na} is low and $[\mathrm{Na}]_{\mathrm{i}}$ is low, the $P_{\mathrm{Na}}[\mathrm{Na}]_{\mathrm{i}}$ term is low and can be ignored because it has only a relatively small effect on the denominator. Similarly, the electrogenic pump term, i.e., $J_{\mathrm{ep}}(RT/FE_m)$ has only a relatively slight effect on the denominator, because the $P_{\mathrm{K}}[\mathrm{K}]_{\mathrm{i}}$ term is large. Therefore, Equation A-4 can be simplified to

$$E_m = \frac{+RT}{F} \ln \frac{P_{\mathrm{K}}[\mathrm{K}]_o + P_{\mathrm{Na}}[\mathrm{Na}]_o + J_{\mathrm{ep}} \frac{RT}{FE_m}}{P_{\mathrm{K}}[\mathrm{K}]_i} \qquad \text{eq. A-12}$$

Further reduction gives

$$\frac{E_m F}{RT} = \ln\left(\frac{[\mathrm{K}]_o}{[\mathrm{K}]_i} + \frac{P_{\mathrm{Na}}[\mathrm{Na}]_o}{P_{\mathrm{K}}[\mathrm{K}]_i} + \frac{J_{\mathrm{ep}}RT}{E_m F P_{\mathrm{K}}[\mathrm{K}]_i}\right) \quad \text{eq A-12a}$$

or

$$e^{E_m F/RT} = \frac{[\mathrm{K}]_o}{[\mathrm{K}]_i} + \frac{P_{\mathrm{Na}}[\mathrm{Na}]_o}{P_{\mathrm{K}}[\mathrm{K}]_i} + \frac{J_{\mathrm{ep}}RT}{E_m F P_{\mathrm{K}}[\mathrm{K}]_i} \quad \text{eq. A-12b}$$

where the right-hand term is the complete electrogenic Na^+ pump term. Because E_m is negative, the pump term is negative and therefore has the effect of making $e^{E_m F/RT}$ smaller (by subtraction); a smaller $e^{E_m F/RT}$ gives a larger numerical resting potential value (V), because

$$e^{E_m F/RT} = e^{-VF/RT} = \frac{1}{e^{VF/RT}}$$

(where the negative sign of E_m is entered separately), and therefore a smaller denominator gives a larger potential; i.e., the cell is hyperpolarized. Moreton's equation allows E_m to be greater, i.e., more negative, than E_K.

Although the Na^+ pump term depends on E_m , it can be shown practically that the pump term can without serious error be regarded as a constant addition to the right side of the equation, provided that the net extrusion of positive charge (J_{ep}) remains constant (Moreton, 1969). Moreton (1969) demonstrated that, provided J_{ep} remains constant, Equation A-12b gives an approximately linear relationship between $e^{E_m F/RT}$ and $[K]_o$ (linear scale) at high values of $[K]_o$; at low $[K]_o$ there is a deviation from linearity in the direction of a decreasing resting potential. The slope of the linear region is equal to $1/[K]_i$; the deviation from the straight line at zero $[K]_o$ gives the contribution to the resting potential from the pump, and the intercept can be used to obtain the ratio of P_{Na}/P_K. In the presence of ouabain, the curve is linear throughout the entire range. Moreton applied his analysis to the snail giant neuron, which exhibits a hyperpolarization averaging about 8 mV as $[K]_o$ is elevated from 0 mM to about 4 mM.

When ouabain is used to block the Na-K pump, the pump term is eliminated, giving

$$e^{(E_m-\Delta E_m)F/RT} = \frac{[K]_o}{[K]_i} + \frac{P_{Na}[Na]_o}{P_K[K]_i} \qquad \text{eq. A-13}$$

where $(E_m-\Delta E_m)$ is the membrane potential measured in the presence of ouabain, and ΔE_m is the electrogenic contribution to the resting potential (amount of depolarization produced immediately by ouabain). Subtracting Equation A-13 from Equation A-12b gives

$$e^{E_mF/RT} - e^{(E_m-\Delta E_m)F/RT} = \frac{J_{ep}RT}{E_mFP_K[K]_i} \qquad \text{eq. A-14}$$

where the exponential term of the nonpumping cell is subtracted from that of the pumping cell. The electrogenic flux can be calculated from Equation A-14: the left side is evaluated from measurements of membrane potential in the absence and presence of ouabain, and P_K can be computed from the following equation [see Christoffersen (1972)]

$$\left(\frac{dI_o}{dE_m}\right)_{I_o\to 0} = G = \frac{F^3P_KE_m\left([K]_o + \frac{P_{Na}}{P_K}[Na]_o\right)[K]_i}{R^2T^2\left([K]_o + \frac{P_{Na}}{P_K}[Na]_o - [K]_i\right)} \qquad \text{eq. A-15}$$

where I_o is the applied current density (amp/cm^2), G is the total membrane specific conductance, and $(dI_o/dE_m)_{Io}{\rightarrow}0$ is the slope of the steady-state current/voltage plot through the origin.

Finally, J_{ep} can also be obtained from Ohm's law

$$V_{ep} = R_m I_{ep} \qquad \text{eq. A-16}$$

or because current (amp/cm^2) equals flux (mol/cm^2 per sec) times F (coul/mol)

$$V_{ep} = R_m F J_{ep} \qquad \text{eq. A-16a}$$

where R_m is the membrane specific resistance (in Ω·cm), and V_{ep} is the electrogenic pump potential contribution to the resting potential. Thus if the net pump current (I_p) is constant, the observed V_{ep} is directly related to R_m. The J_{ep} value calculated from Equation A-16a is close to that obtained from Equation A-14 for snail neurons (Luttgau & Glitsch, 1976). Rapoport (1970) demonstrated that the use of Ohm's law is an oversimplification and that its use should be corrected for the change in diffusion potential associated with voltage-dependent conductance changes. This correction, however, is negligible when the pump potential contribution is slight, e.g., < 8 mV (Luttgau & Glitsch, 1976).

Chapter X

DEVELOPMENTAL CHANGES IN K^+ PERMEABILITY

Important changes — including ultrastructural, metabolic, pharmacological, and electrophysiological changes — occur in myocardial cells during embryonic development (Sperelakis, 1972a; Sperelakis & McLean, 1978a, b). In this chapter, we examine some of the striking changes that occur in the electrical properties of ventricular myocardial cells during embryonic development of chick hearts. The electrical properties at each stage of development not only determine many of the functional properties of the heart at that stage but also give important clues for analyzing the complex electrophysiology of adult hearts.

The tubular heart of the chick embryo begins contracting spontaneously and propelling blood at 30–40 h; the blood pressure is extremely low (1–2 mmHg) at this stage (Romanoff, 1960). As measured by intracellular microelectrodes, the transmembrane resting potential (E_m) of the ventricular portion of the tubular chick heart is about -40 mV on embryonic day 2, and the resting potential increases to about -51 mV on day 3 (Table X-1; Sperelakis & Shigenobu, 1972). The greatest increases in resting potential occur between days 2 and 7. The resting potential is close to -80 mV by day 11, nearly the final adult value. As is discussed later in this subsection, the great increase in resting E_m during the first few days may be due mainly to an increase in P_K rather than in E_K. Although it is possible that low recorded potentials in young hearts can be caused by improper sealing of the microelectrode (current leakage around the electrode tip is greater in cells having a high input resistance), similar low resting potentials in young hearts have been reported (Lehmkuhl & Sperelakis, 1963; Shimizu &

Tasaki, 1965; Yeh & Hoffmann, 1967; Pappano, 1972; Bernard, 1975).

The relationship between resting potential and $[K]_o$ is illustrated for three hearts of different ages (data points) in Figure X-1. Also plotted are theoretical curves calculated from the Goldman constant field equation, given in the inset of the figure; the curves are for five different ratios of P_{Na}/P_K: 0.001, 0.01, 0.05, 0.1, and 0.2. For these calculations it was assumed that $[Na]_i$ was 15 mM and $[K]_i$ was 150 mM (estimated by extrapolation of the curves connecting the data points to zero potential). It is seen that the data points for the 3-day-old heart most closely fit the theoretical curve for a P_{Na}/P_K ratio of 0.2; those for the 5-day-old heart fit the curve for a P_{Na}/P_K ratio of 0.1; and those for the 15-day-old heart most closely fit between the 0.05 and 0.01 curves. These results suggest that the P_{Na}/P_K ratio is high in young hearts and that this accounts for the low resting potential; that is, the low resting potential is not due to a low $[K]_i$ (hence a low E_K). As given in Table X-1, $[K]_i$ is already substantially high (125 mM, in 2-day-old hearts); a maximal value of 155 mM is achieved by day 14. Only a small increase in the calculated E_K occurs during development: from about -90 mV on day 2 to -94 mV on days 7–20 (Table X-1).

Thus, in young hearts the resting potential is far from E_K , because of the high P_{Na}/P_K ratio. There has been a similar finding in adult sinoatrial nodal cells (H. Irisawa and A. Noma, personal communication). In old embryonic or adult hearts the E_m vs. log $[K]_o$ curve is nearly linear at more than 10 mM K^+, with a slope approaching the theoretical 60 mV/decade (from the Nernst equation) and indicating that E_m is virtually completely determined by E_K. The slope for hearts 7–20 days old is 51–53 mV/decade, whereas that for 4-day-old, 3-day-old, and 2-day-old hearts is 46, 40, and 30 mV/decade, respectively (Fig. X-1 and Table X-1). The curves for hearts with high

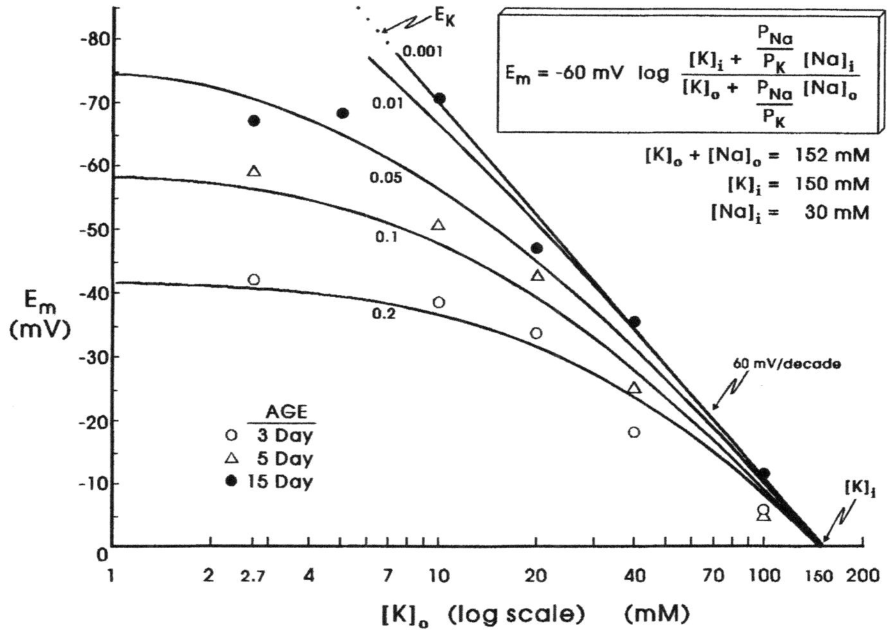

FIGURE X-1. Resting potential (E_m) as a function of $[K]_o$, for chick embryonic hearts 3 days old (open circles), 5 days old (closed triangles), and 15 days old (closed circles). Curves represent calculations from Goldman constant field equation (inset) for P_{Na}/P_K ratios of 0.001, 0.01, 0.05, 0.1, and 0.2. These calculations were made for a $[K]_i$ of 150 mM and a $[Na]_i$ of 30 mM. Sum of $[K]_o$, and $[Na]_o$ was held constant at 152 mM; i.e., as $[K]_o$ was elevated, $[Na]_o$ was lowered by an equimolar amount. Broken straight line gives potential of K^+ (E_K) calculated from the Nernst equation and nearly superimposes over the curve for a P_{Na}/P_K of 0.001. Data for 3-day-old hearts most closely fit the theoretical curve for a P_{Na}/P_K ratio of 0.2, data for the 5-day-old hearts fit closest to a ratio of 0.1, and those for the 15-day-old hearts fit between the curves for a ratio of 0.05 and 0.01. $[K]_i$ and $[Na]_i$, intracellular K^+ and Na^+ concentrations, respectively; $[K]_o$ and $[Na]_o$, extracellular K^+ and Na^+ concentration, respectively. [From Sperelakis and Shigenobu (1972).]

TABLE X-1

Summary of data obtained from resting potential vs. log [K] curves for ventricular cells from chick embryonic hearts at various stages of development.

Embryonic Age days	E_m mV	Slope mV/decade	$[K^+]_i$ mM	E_K mV	Average $\frac{P_{Na}}{P_K}$ Ratio	r_{in} MΩ
2	-40	30	125	-90	0.21	13.0
3	-51	40	130	-91	0.17	8.5
4	-57	46	140	-93	0.08	6.5
5-6	-58	50	130	-91	0.08	5.5
7-9	-71	51	145	-94	0.07	5.5
11-13	-80	53	145	-94	0.07	4.7
14-20	-78	52	155	-95	0.05	4.5

r_{in} = input resistance. E_m and E_K values are given for a $[K]_o$ of 4.0 mM. The slope is the average at $[K]_o$ levels between 10 and 100 mM. $[K]_i$ was estimated from the extrapolation of fitted linear curves to zero potential. [Data from Sperelakis and Shigenobu (1972) and from Sperelakis et al (1975).] Similar values for slope and $[K]_i$ were obtained by Pappano (1972) for embryonic chick atrium at 4 days, 6 days, and 12 days, but the 18-day values were -59 mV/decade and 125 mM $[K]_i$.

TABLE X-2

Influence of extracellular $[K^+]$ on K^+ permeability of chick embryonic hearts.

	P_K cm/sec (x 10^7)		P_{Na} cm/sec (x 10^7)		$\frac{P_{Na}}{P_K}$	
$[K]_o$ mM	6-8 days	18-20 days	6-8 days	18-20 days	6-8 days	18-20 days
0	0.33	0.95	0.044	0.063	0.133	0.066
2.5	1.32	3.10	0.042	0.053	0.032	0.017
5.0	1.32	2.40	0.057	0.046	0.043	0.019
10	1.44	2.42				
20	1.39	2.65				

Values for P_{Na} were calculated from measured P_K values (from flux studies) and measured P_{Na}/P_K ratios (obtained from the constant-field equation). The flux density values, on which the P_K calculations are based, were obtained from measured volume/surface area ratios (by a stereologic method) of 1.24 x 10^{-4} cm for both 7-day-old and 19-day-old hearts. The measured $[K^+]_i$ values were 151 mM (6-8 days) and 122 mM (18-20 days), and the corresponding $[Na^+]_i$ values were 16 mM and 15 mM. [Data from Carmeliet et al. 1976.]

P_{Na}/P_K ratios continually bend, and so the values just given are for the average slope. Pappano (1972) found similar values for $[K^+]_i$ and a similar slope for embryonic chick atrial cells at various stages of development. For example, he reported a $[K]_i$ of 145 mM on day 4. Using electrolyte analyses, Carmeliet and associates (1976) reported $[K]_i$ values of about 150 mM for 7-day-old chick hearts. Thus there is little doubt that $[K]_i$ is already high in young hearts. Hence the cardiac cells must actively transport cations before day 2, and even during the precardiac blastoderm stage. [Strangely enough, the $[K]_i$ level may actually be somewhat lower on days 18–20 than it is on days 6–8 (Harsch & Green, 1963; Carmeliet et al., 1976). In contrast to these results, it was found that the internal K^+ activity increased by 17% between day 4–5 and day 7–8, but the activity remained constant thereafter (H.A. Fozzard, personal communication).]

To help determine whether the P_{Na}/P_K ratio is high in young hearts because of a high P_{Na} or because of a low P_K, input resistance was measured from steadystate voltage-current curves (Sperelakis & Shigenobu, 1972). The input resistance (r_{in}) of the ventricular cells is high (13 MΩ) in young 2-day-old hearts and rapidly declines during the next few days, r_{in} reaching the final adult value (4.5 MΩ) by day 12 (Table X-1). If the average cell size and the electrical arrangement of the cells remains unchanged during this period, the high r_{in} of the young hearts suggests that membrane resistivity (R_m) is high. The latter would be consistent with a low g_K and P_K in young hearts. Consistent with a low P_K in young hearts is the finding that the chronaxie of 2-day-old hearts is about fourfold greater than that of 9- to 16-day-old hearts (Shimizu & Tasaki, 1965); this greater chronaxie suggests that the membrane time constant, and hence membrane resistivity, is about fourfold greater in young hearts (assuming capacitance remains constant). On the basis of ^{42}K flux measurements, Carmeliet et al. (1976) also have reported that P_K is about two- to threefold less in 6- to 8-day-old hearts than in 18- to 20-day-old hearts (Table X-2); this finding is consistent with the conclusions from the electrical studies described earlier in this paragraph. Their P_{Na} calculations suggested that this parameter did not change greatly during development. Although Carmeliet et al. (1976) reported that the calculated P_K for 3- to 5-day-old hearts was nearly as great as that for the 18- to 20-day-old hearts (which disagrees with the electrical measurements), they did not control for the spontaneous beating and action potentials in the young hearts; the increased K^+ efflux would result in an erroneously high calculation for P_K.

As expected from the greater flattening of the E_m vs. log $[K]_o$ curve at lower $[K]_o$ levels, i.e., less depolarization for a given increment in $[K]_o$, young hearts are less affected by an elevation of $[K]_o$ than are older hearts (Sperelakis & Shigenobu, 1972). Both the inhibition of automaticity of the whole heart as well as the loss of ventricular excitability are less affected. The action potential of old embryonic chick atrial muscle is markedly shortened by acetylcholine, and the cells are slightly hyperpolarized because of an increase in g_K. The atrial cells of young hearts, however, are slightly depolarized by acetylcholine in a normal medium (Pappano, 1972), even though a large hyperpolarization is theoretically possible because E_m is much below E_K. Because the cells were slightly hyperpolarized in Na^+-free medium, Pappano (1972) suggested that acetylcholine increases both g_{Na} and g_K in the atrial muscle of young hearts and that the cells have a low P_K (namely, that few K^+ channels are available to be opened by acetylcholine).

As would be predicted from the changes in P_K, great changes in automaticity of the ventricular cells also occur during development. The major requirements for automaticity appear to be a low g_{Cl} , as is generally true for myocardial cells, and a low g_K. (The low g_K enhances membrane inductance in series with the negative slope conductance of one type of K^+ channel and tends to cause oscillations in membrane potential. The low g_K also produces some depolarization, i.e., moves E_m farther from E_K , and places the membrane potential in the region that can support pacemaker oscillations.) The incidence of hyperpolarizing afterpotentials and pacemaker potentials is high (80–100%) in young hearts, and this incidence decreases to 0% in old embryonic hearts (Sperelakis & Shigenobu, 1972). If a portion of the ventricle is cut and isolated to remove drive from the nodal cells, the incidence of pacemaker potentials observed in the impaled cells is 100% for embryos up through day 10, whereas the incidence is 0% in embryos 12 days old or older. These results indicate that the ventricular myocardial cells possess automaticity capability when they are young but that this capability diminishes as the cells become older because of the increase in P_K , The old ventricular cells, however, again become automatic, a change associated with a decrease in g_K , when they are trypsin dispersed and placed into cell culture (Sperelakis, 1972a).

The specific activity of the (Na-K)-ATPase is low in young embryonic chick hearts and increases during development (Sperelakis, 1972b). The average value of the activity on day 4 is about 35% of that on day 16, that on day 6 is about 42%, and that on day 13 is about 77%. The ATPase activity on day 16 is about equal to the adult level. Thus while P_K is increasing during development — and hence there is an increase in the outward passive leak of K^+ and the inward leak of Na^+ (due to the increased electrochemical driving force) — the capability of the Na^+-K^+ pump is increasing; that is, the increased cation pump capability tends to compensate for the increased demand on the pump, a demand that is due to an increase in the cation leak and an increase in the heart rate. The pumping capacity of the young hearts, however, must be sufficient to maintain the high $[K]_i$ and low $[Na]_i$ already present.

Electrolyte analyses of chick embryonic hearts (ventricles) indicate that the total tissue content of Na^+ in young hearts is exceedingly high and that Na^+ decreases gradually until about day 7, after which the level remains constant (Harsch & Green, 1963; Klein, 1960; Yeh & Hoffmann, 1967). Much of this Na^+ in young hearts, however, is bound to extracellular (subendocardial) mucopolysaccharide cardiac jelly (Thureson-Klein & Klein, 1971). The data from electrophysiological studies indicates that the thermodynamically active free intracellular Na^+ must not be very high, because the Na^+-dependent action potentials overshoot to +11 mV in day 2 hearts; the overshoot increases rapidly and reaches the adult value of +28 mV by day 7 (Sperelakis & Shigenobu, 1972). Carmeliet and associates (Carmeliet et al., 1976) measured $[Na]_i$ values of about 15 mM for both 6- to 8-day-old and 18- to 20-day-old chick hearts. Thus there seems little doubt that the free $[Na]_i$ is already low by day 6.

The cyclic AMP level is relatively high in young hearts and decreases during development (McLean et al., 1975; Thakkar & Sperelakis, 1987). In 4-day-old hearts, the cyclic AMP level is 116 pmol/mg protein, and by day 5 this level decreases sharply to about 41. There is a gradual further decline until the adult level of about 10 pmol/mg protein is reached by about day 16. The relationship, if any, between changes in membrane properties and changes in cyclic AMP levels during development of heart remains to be clarified. In cultured skeletal muscle the cyclic AMP level decreases sharply as the myoblasts fuse into myotubes, i.e., when the cells further differentiate (Reporter, 1972). In adult or old embryonic hearts an increase in the cyclic AMP level is associated with an increase in the number of available slow Ca^{2+} channels (Shigenobu & Sperelakis, 1972). The cyclic GMP level is very low in young embryonic chick hearts, and progressively increases during development (Thakkar & Sperelakis, 1987).

The action potentials of the cells of intact chick hearts also undergo sequential changes during development in situ (Sperelakis & Shigenobu, 1972). In brief, the young embryonic hearts fire action potentials whose inward current is carried mainly by Na^+ through slow Na^+ channels that are insensitive to TTX and that inactivate at a much lower voltage. Fast Na^+ channels that are sensitive to TTX make their initial appearance on about day 4 or 5 of development and increase in density thereafter.

The changes that occur in membrane electrical properties during cell culture of myocardial cells, including changes in P_K , have been reviewed (Sperelakis, 1972a; Sperelakis et al., 1973; Sperelakis & McLean, 1978b).

Chapter XI

GIBBS-DONNAN POTENTIALS

Because intracellular cytoplasm contains many colloids, including large nondiffusible polyvalent electrolytes, the possibility exists of establishing a Donnan equilibrium across the cell membrane with an accompanying transmembrane Gibbs-Donnan potential. The resting potential of most cells in the body, including nerve and muscle cells, however, is not due to a Donnan equilibrium, and the normal resting potential is not a Gibbs-Donnan potential, as is erroneously stated in some textbooks. In the true Donnan equilibrium, all diffusible ions are in equilibrium across the membrane. But many ions — like Na^+, K^+, Ca^{2+}, and H^+ — in nerve and muscle cells are not in equilibrium; that is, $E_{Na} \neq E_m$, $E_K \neq E_m$, $E_{Ca} \neq E_m$, and $E_H \neq E_m$. On the other hand, Cl^- is at equilibrium (i.e., passively distributed) in most vertebrate cells; namely, $E_{Cl} = E_m$ In addition, a large internal pressure and concomitant swelling (of animal cells) would occur if a Donnan equilibrium were allowed to become established. It is the action of one or more cation pumps that keeps the Donnan osmotic pressure from developing and keeps certain cations out of equilibrium. Thus a second important function of the Na-K pump is the regulation of cell volume. Inhibition of active ion transport by any means leads to osmotic swelling due to the establishment of the Donnan equilibrium. Under such conditions the cells gain Na^+, Cl^-, H_2O, Ca^{2+}, and they lose K^+. Thus the Gibbs-Donnan potential is passive; i.e., energy is not necessary to its establishment. In contrast, the resting potential is actively generated (indirectly or directly) by the action of the Na^+-K^+ pump. The Gibbs-Donnan potential is usually less than -20 mV, whereas the resting potential is -40 to -100 mV, depending on the cell (and its ratio of P_{Na} to P_K).

For the Donnan equilibrium, note that a small membrane potential is established even though the artificial membrane used in a laboratory experiment may be equally permeable to the small diffusible ions used (Fig. XI-1). For example, if aqueous solutions of 0.1 M $(Na^+)_n$-proteinate^{-n} (side 1) and 0.1 M NaCl (side 2) are initially placed on the two sides of a chamber separated by a collodion membrane, and if $g_{Na} = g_{Cl}$, then at equilibrium $E_{Na} = E_{Cl} = -18$ mV; the side containing the protein anion becomes negative with respect to the other side. Thus because both diffusion potentials have the same polarity (as well as magnitude), a PD occurs across the membrane, even though conductances for Na^+ and Cl^- across the membrane may be equal. The osmotic pressure of the solution on side 1 containing the nonpermeant protein is greater than that on side 2. A more complete explanation for the development of the Gibbs-Donnan potential follows.

The Gibbs-Donnan potential (which is an equilibrium PD) does not depend on metabolic energy. This potential is passively produced by the concentration gradients for diffusible electrolytes (e.g., Na^+ and Cl^-) across a membrane, gradients caused by the presence of one or more large nondiffusible (with respect to the membrane) polyvalent electrolytes (e.g., negatively charged proteins) on one side of the membrane. In essence, the negatively-charged protein molecules (at pH 7) inside the cell bind cations (e.g., Na^+ or K^+) and repel anions (e.g., Cl^-). Therefore in the Gibbs-Donnan situation, the inside of the cell has a higher concentration of Na^+ (or K^+) and a lower concentration of Cl^- than does the solution bathing the cell. The equilibrium potentials for Na^+ (E_{Na}) and for Cl^- (E_{Cl}) are equal in magnitude and are of the same sign, thereby producing a PD across the membrane. The PD is negative on the inside (side containing the protein) and usually is about -20 mV or less. This discussion applies to a cell that either has no

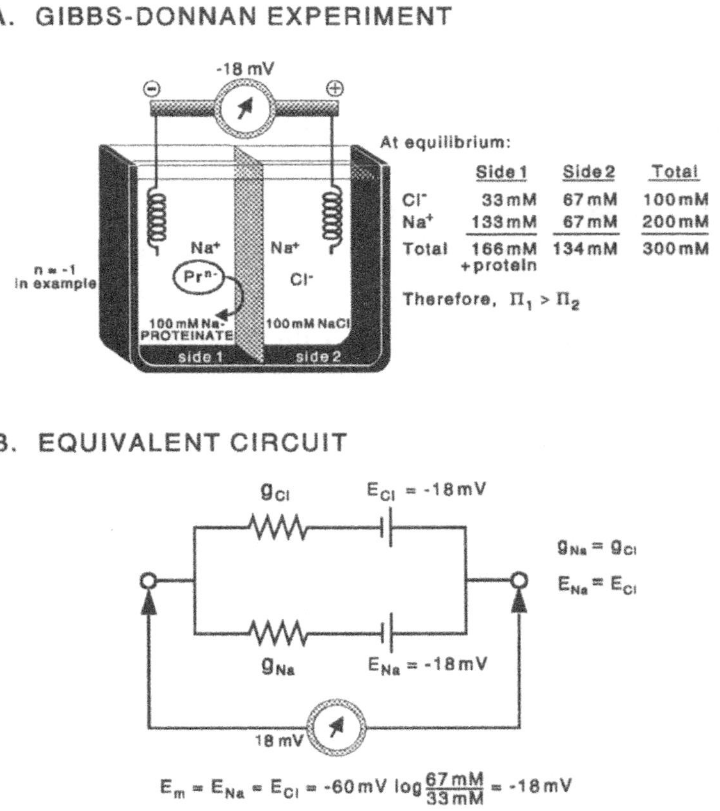

FIGURE XI-1. Gibbs-Donnan potential. **A:** Diagram depicts the experimental arrangement for obtaining Gibbs-Donnan potential. A membrane freely permeable to all small ions, but impermeable to the large protein molecules, is used to separate two solutions, only one of which (side 1) contains protein. Side containing the protein becomes negative, with respect to the other side, by a small voltage (-18 mV in example). This membrane potential (E_m) does not depend on active ion transport or on selective permeability properties of the membrane, as normal cell resting potential does. The diffusible ions (Na^+ and Cl^- in example), however, become unequally distributed across the membrane, and it is their diffusion potentials ($E_{Na} = E_{Cl}$) that produce the Gibbs-Donnan potential. **B:** Equivalent circuit for Gibbs-Donnan experiment (depicted in A) demonstrating that $E_m = E_{Na} = E_{Cl}$, the Na^+ and Cl^- batteries being of equal magnitude and of the same sign Therefore the relative conductances of the membrane to Na^+ and Cl^-, whether equal or not, are irrelevant to the potential. g_{Cl} and g_{Na} conductance for Cl^- and Na^+, respectively.

ATP for pumping ions against electrochemical gradients or that has its Na-K pump completely blocked by either ouabain or another agent.

To quantitate the ion distributions produced and the PD developed, let us examine the artificial system shown in Figure XI-1A. In this system, a chamber was separated into two compartments by a collodion membrane, which has small uncharged pores that allow Na^+ and Cl^- ions, but not large protein molecules, to diffuse through. A 100-mM solution of Na-proteinate was added to one side (compartment 1), and a 100-mM solution of NaCl to the other side (compartment 2). An electrode was positioned on each side so that the PD across the membrane (37° C) could be recorded. Let us presume that the Na-proteinate was completely ionized and, for simplicity, that the protein had a net negative charge of only one. Thus there was at the first instant no diffusion force for Na^+, but there was for Cl^-, since Cl^- is 100 mM in compartment 2 and 0 mM in compartment 1. But because the principle of electroneutrality in the bulk solution cannot be violated (i.e., there must be an equal number of cations and anions), Na^+ must accompany the diffusion of Cl^- from side 2 to side 1. (In actuality, there is a small charge separation directly across the membrane to account for the PD, side 2 of the membrane having a small excess of Na^+ ions and side 1 a small excess of Cl^- ions.) So one thing that must be true when the system comes to equilibrium is that

$$[Na^+]_2 = [Cl^-]_2 \qquad \text{eq. 1}$$

Furthermore, the increase in Na^+ on side 1 must be exactly equal to the increase in Cl^- on side 1. Thus the concentration difference of Na^+ that is built up at equilibrium must be exactly equal to the final concentration difference for Cl^-. This is because the large initial gradient for Cl^- is what drives the Na^+ to make its gradient. Therefore, it must also be true that

$$\frac{[Na^+]_1}{[Na^+]_2} = \frac{[Cl^-]_2}{[Cl^-]_1} \qquad \text{eq. 2}$$

Cross multiplying gives

$$[Na^+]_1[Cl^-]_1 = [Na^+]_2[Cl^-]_2 \qquad \text{eq. 2a}$$

Another way to consider this is that E_{Na} must equal E_{Cl} , and therefore

$$E_{Na} = E_{Cl} \qquad \text{eq. 3}$$

$$\frac{-60\text{ mV}}{+1}\log\frac{[Na^+]_1}{[Na^+]_2} = \frac{-60\text{ mV}}{-1}\log\frac{[Cl^-]_1}{[Cl^-]_2} \qquad \text{eq. 3a}$$

$$= \frac{-60\text{ mV}}{+1}\log\frac{[Cl^-]_2}{[Cl^-]_1}$$

Dividing both sides by -60 mV and removing the log gives Equation 2:

$$\frac{[Na^+]_1}{[Na^+]_2} = \frac{[Cl^-]_1}{[Cl^-]_2} \qquad \text{eq. 2}$$

Equation 2a indicates that the product of the diffusible ions on side 1 must be equal to the product of the diffusible ions on side 2. From the Nernst equation, the following relationships can be given:

$$E_{Na} = \frac{-RT}{zF} \ln \frac{[Na]_1}{[Na]_2} \qquad \text{eq. 4}$$

$$E_{\text{Na}} = \frac{-61 \text{ mV}}{+1} \log \frac{[\text{Na}^+]_1}{[\text{Na}^+]_2} \qquad \text{eq. 4a}$$

and

$$E_{\text{Cl}} = \frac{-61 \text{ mV}}{-1} \log \frac{[\text{Cl}^-]_1}{[\text{Cl}^-]_2} \qquad \text{eq. 5}$$

because Cl^- is negative (z = -1), whereas Na^+ is positive (z = +1). Equation 5 is the same as (note that a negative sign in front of a log inverts the ratio)

$$E_{\text{Cl}} = -61 \text{ mV} \log \frac{[\text{Cl}^-]_2}{[\text{Cl}^-]_1} \qquad \text{eq. 5a}$$

Because $[\text{Na}]_1/[\text{Na}]_2 = [\text{Cl}]_2/[\text{Cl}]_1$, as Equation 2 indicates, and from Equations 4a and 5a, it is clear that Equation 3 holds true:

$$E_{\text{Na}} = E_{\text{Cl}} \qquad \text{eq. 3}$$

For quantitation, let us use x to indicate the amount (in mM) of Cl^- or Na^+ that shifted from side 2 to side 1 at equilibrium. Then the amount of Na^+ on side 2 is 100 mM – x (the original amount minus the amount lost); Cl^- on side 2 is also 100 mM – x, since $[\text{Na}]_2 = [\text{Cl}]_2$. The Na^+ on side 1 at equilibrium is 100 mM + x (the original amount plus the amount gained), and the Cl^- on side 1 is simply x. Therefore, by substituting these values into Equation 2a, the value for x can be obtained:

$$[Na^+]_1[Cl^-]_1 = [Na^+]_2[Cl^-]_2 \qquad \text{eq. 2a}$$
$$(100+x)x = (100-x)(100-x)$$
$$100x + x^2 = 10{,}000 - 200x + x^2$$
$$300x = 10{,}000$$
$$x = 33.3$$

Thus, at equilibrium $[Cl]_1$ = 33 mM; $[Na]_1$ = (100 + 33) = 133 mM; $[Cl^-]_2$ = (100 – 33) = 67 mM; $[Na]_2$ = (100 – 33) = 67 mM. These values are summarized in Figure XI-1A. Note that all the equations and conditions are obeyed. The Gibbs-Donnan potential produced then is

$$E_{Na} = \frac{-60\text{ mV}}{+1}\log\frac{133\text{ mM}}{67\text{ mM}} \qquad \text{eq. 6}$$
$$= -18\text{ mV}$$

and

$$E_{Cl} = \frac{-60\text{ mV}}{-1}\log\frac{33\text{ mM}}{67\text{ mM}} \qquad \text{eq. 7}$$
$$= -18\text{ mV}$$

Hence, $E_{Na} = E_{Cl}$; i.e., the two diffusion potentials are equal in magnitude and are of the same sign. Therefore, relative permeability of the membrane to Na^+ and Cl^- is irrelevant, and the PD across the membrane is -18 mV; side 1 containing the protein is negative. The equivalent circuit for this example at equilibrium is given in Figure XI-1B.

Finally, we should note that at equilibrium the sum of Na^+ and Cl^- on side 1 (166 mM) is greater than that on side 2 (134 mM). In addition, there is 100 mM protein on side 1. Thus the total osmotic concentration on side 1 is 266 mosM, compared to 134 mosM on side 2. Therefore there is a large osmotic gradient between the two sides, so water moves from side 2 to side 1 (i.e., water accompanies the net movement of Na^+ and Cl^-) until the hydrostatic pressure head buildup is sufficient to oppose further net movement of water. (The biological cell swells when a Gibbs-Donnan equilibrium is allowed to develop following a blockade of active ion transport for long periods.) The situation is actually more complex because the net water movement into side 1 acts to dilute the ion concentrations building up there, and therefore a true Gibbs-Donnan equilibrium can only become established if the net water movement is stopped, i.e., by allowing an osmotic pressure gradient to develop by making side 1 a closed, or rigid, system. Otherwise, theoretically all of the water moves out of side 2. The osmotic pressure (Π, in atm) of each solution is equal to the osmotic concentration in osmol/L (C) times the isotonic coefficient (i) times the gas constant (R, 0.082 liter-atm/mol-K) times the absolute temperature (T, in K); i.e., $\Pi = iCRT$.

Chapter XII

EFFECT OF RESTING POTENTIAL ON RATE OF RISE OF ACTION POTENTIAL

The greater the magnitude of the resting potential (E_m), the greater the maximum rate of rise of the action potential (max *dV/dt*) up to some maximal value (Fig. XII-1B). The max *dV/dt* is usually close to its maximal value at the normal resting potential (about -80 mV for myocardial cells) but decreases markedly with small depolarizations, as shown in the figure. The max *dV/dt* goes to zero at about -50 mV; i. e., membrane excitability is lost. The reason for this behavior is given by the E_m vs. h_∞ curve (Fig. XII-1A). In Hodgkin-Huxley notation, *h* is the inactivation variable for the fast Na^+ conductance; it is a dimensionless probability factor that varies between 0 and 1.0, and it deals with the open ($h = 1.0$) vs. closed ($h = 0$) positions of the inactivation gate of the channel. The value of *h* is a function of E_m, and h_∞ is the *h* value at steady state, i.e., $t = \infty$ (practically, $t > 20$ msec). As illustrated in Figure XII-1A, the h_∞ curve is sigmoidal, h_∞ being nearly 1.0 at the resting potential (-80 mV) and rapidly diminishing to zero at about -50 mV. Thus, the change in h_∞ as a function of E_m is the cause of the change in max *dV/dt* as a function of E_m.

Thus, partial depolarization of the membrane for whatever reason (e.g., small elevation in $[K]_o$) has profound effects on the action potential rate of rise, and therefore on the propagation velocity. Changes in propagation velocity — whether in nerve or muscle — of course have pronounced effects on physiological function. Conversely if the membrane is significantly hyperpolarized by whatever mechanism, e.g., by an abnormally low $[K]_o$ or by stimulation of an electrogenic pump, this hyperpolarization also can have profound effects on

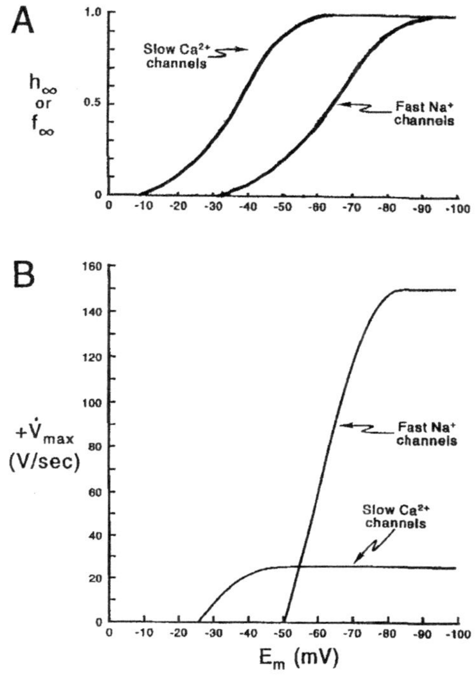

FIGURE XII-1. Graphic representation of differences in behavior, with respect to voltage inactivation, of the fast Na^+ channels and slow (Na^+ and Ca^{2+}) channels. **A:** h_∞ inactivation factor of Hodgkin-Huxley ($g_{Na} = \bar{g}_{Na} m^3 h$ where g_{Na} is the Na^+ conductance, m and h are variables, and overbar means maximal) for fast Na^+ channels or the comparable f_∞ factor for slow channels as a function of resting membrane potential (E_m). The h_∞ or f_∞ represents h or f at t = infinity or steady state (practically, after 20 msec for h and after 1 sec for f). This graph illustrates that fast Na^+ channels of myocardial cells begin to inactivate at about -75 mV and complete inactivation occurs at about -50 mV ($h_\infty = 0$). In contrast, slow channels inactivate between about -45 mV (f_∞ about 1.0) and -25 mV ($f_\infty = 0$). **B:** Maximal rate of rise of the action potential (max dV/dt) as a function of resting E_m for the normal cardiac action potential (dependent on inward current through the fast Na^+ channels) and for the slow action potential (or slow response, dependent on inward current through the slow channels) elicited in cells whose fast Na^+ channels are blocked (by TTX or by depolarization to about -45 mV). max dV/dt is a measure of the inward current intensity (everything else, such as membrane capacitance, held constant), which in turn dependent on the number of channels available for activation.

function. As shown in Figure XII-1, hyperpolarization does not greatly increase h_∞ or max dV/dt. Hyperpolarization does, however, increase the amount of depolarization (the so-called critical depolarization) required for any point in the membrane to reach the threshold potential during the conduction of an action potential. Because the local-circuit current available to bring each point in the membrane to threshold is limited (dependent on max dV/dt at a fixed C_m), a sufficient hyperpolarization can lead to a loss of conduction.

The kinetically slow cationic channels also have an inactivation gate, and therefore the equivalent of an h variable (f variable). The main difference is that the inactivation curve is shifted to the left, so that inactivation begins at -45 mV and is complete at about -25 mV (Fig. XII-1A). Slow channels, compared to fast Na^+ channels, also have a lower activation (threshold) potential (about -35 mV vs. -60 mV), and they turn on (activate) more slowly, turn off (inactivate) more slowly, and recover more slowly. In myocardial cells if the fast Na^+ channels either are blocked by TTX or are voltage-inactivated in elevated (25 mM) $[K]_o$ (E_m of -40 mV), the addition of some positive inotropic agents — such as catecholamines, histamine, angiotensin, fluoride ion, methylxanthines, or cyclic AMP — restores excitability in the form of slowly rising but overshooting action potentials, referred to as the slow response (Shigenobu & Sperelakis, 1972; Schneider & Sperelakis, 1974, 1975). These agents presumably increase the number of available slow channels in the membrane, and the action of many of these agents is accompanied by an increase in the level of intracellular cyclic AMP. Loss of the slow responses does not occur until the membrane is depolarized to about -25 mV (Fig. XII-1B).

CHAPTER XIII

CABLE PROPERTIES AND PROPAGATION

A. Introduction

Nerve and muscle fibers make relatively poor cables, in comparison with commercial cables (such as a transoceanic submarine cable) (Sperelakis, 1993). There would be a great deal of energy loss in the transmission of information along these biological cables if the cable remained passive (no active impulse). The signal would become greatly attenuated and distorted after traveling only a relatively short distance, e.g., 1 mm. Hence, biological signal transmission in long fibers is an active energy-consuming process, the energy input being distributed along the entire length of the fiber. The biological cable is poor for the following reasons: (a) R_i is approximately 10^7 times that for copper wire. (b) The C_m is high. (c) R_m is approximately 10^{-6} times that for an effective insulator such as rubber. (But the true resistivity of the cell membrane is comparable to that of an efficient commercial insulator; that is, the cell membrane would be as efficient as a commercial insulator if both were equally thick.) The biological cable, however, is the best possible, given the biological circumstances and requirements. In addition, the nerve cable has been vastly improved by an evolutionary development in the vertebrates (namely, myelination); myelination increases R_m and decreases C_m by more than 100-fold. There are two consequences that result from improvement of the nerve cable by myelination: (a) the velocity of propagation is greatly increased, and (b) the energy cost of signaling is greatly decreased.

It is imperative that the body be able to transmit a signal from one point to another very rapidly. The only way that this can be accomplished is by an electrical mechanism. Blood flow and diffusion of signalling molecules are much too slow to allow rapid signalling. In contrast, electricity flows very quickly, at the speed of light (300,000,000 m/sec) in a copper wire or about 1/9th the speed of light in a water solution (like the body is made up of). Therefore, the body makes use of electricity for rapid signalling in the nervous system, skeletal muscle, heart, and smooth muscles. Propagation velocity in our fastest nerves is about 120 m/sec, about 6 m/sec in skeletal muscle, about 1.0 m/sec in heart, and about 5 cm/sec in smooth muscle (Tables I-1 and XIII-1).

As one example of the need for very fast communication or signalling, let's look at the process of walking. Very rapid signals must travel from the motor cortex of the brain, down to the lower spinal cord region, and out the motor axons to the skeletal muscles of the lower extremities (Fig. XIII-1). In this process, the signal crosses one or more synapses, which are regions in which one neuron ends and the next one begins and in which a special chemical neurotransmitter signal is involved (Fig. XIII-2). At the termination of each branch of a motor nerve axon on the skeletal muscle fiber, there is another synapse, namely the neuromuscular junction or motor end-plate (Fig. XIII-1). The signal crosses the neuromuscular junction, and gives rise to an AP in the muscle fiber that propagates in both directions from the motor end-plate. The muscle AP elicits contraction. Receptors in the muscles (e.g., stretch receptors) transmit information (in the form of propagating APs) back into the central nervous system (CNS). Thus, in walking, there is a continual rapid flow of information and instructions to the muscles in both directions: out of the CNS and into the CNS. Therefore, even a relatively simple skeletal activity as walking would not be possible without a very rapid signalling system. To drive home this last point, in various demyelinating diseases (e.g., caused by some viruses, heavy metals, autoimmune reactions), loss of the myelin sheath

Table XIII-1

Conduction Velocity as a Function of Fiber Diameter in Various Nerve Axons and Muscle Fibers

Fiber Type	Fiber Diameter (μm)	Propagation Velocity (m/sec)	Velocity/Diameter (m/sec/μm)
Myelinated axons	20	120	6.0
	12	75	6.2
	5	30	6.0
Nonmyelinated axons	1.5	2.0	1.3
	1.0	1.3	1.3
Squid giant axons (20° C)	500	25	0.05
Skeletal muscle fibers	50	6.0	0.12
Cardiac muscle fibers	15	0.5	0.03
Smooth muscle fibers	5	0.05	0.01

Reproduced from Table 1 in Sperelakis, N., 1993 (Cable Properties and Propagation. Ch 5, pp. 83-97; In Essentials of Physiology (N. Sperelakis & R.O. Banks, eds.), Little Brown and Co., Boston, 1993.)

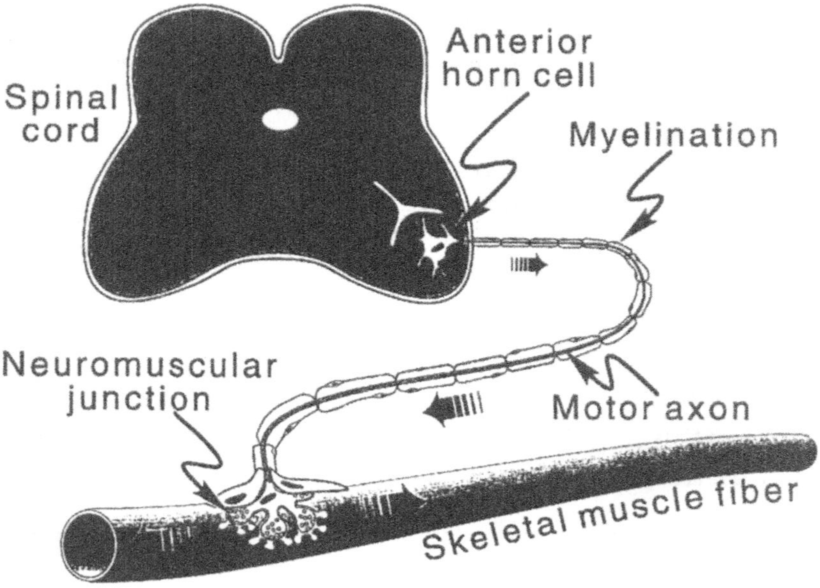

Figure XIII-1. Schematic diagram of a motor axon, with the cell body (soma) in the anterior horn of the spinal cord and its terminal branches ending on skeletal muscle fibers (only one muscle fiber depicted) to form the neuromuscular junctions (motor end-plates). Each motor axon with its attached skeletal muscle fibers is known as the motor unit. The motor axons are of large diameter (e.g., 20 μm) and are myelinated, and therefore propagate at fast velocities (e.g., 120 m/sec). As a consequence of the chemical synaptic transmission process at the neuromuscular junction, an action potential is initiated in the muscle fiber, and propagates in both directions and brings about contraction.

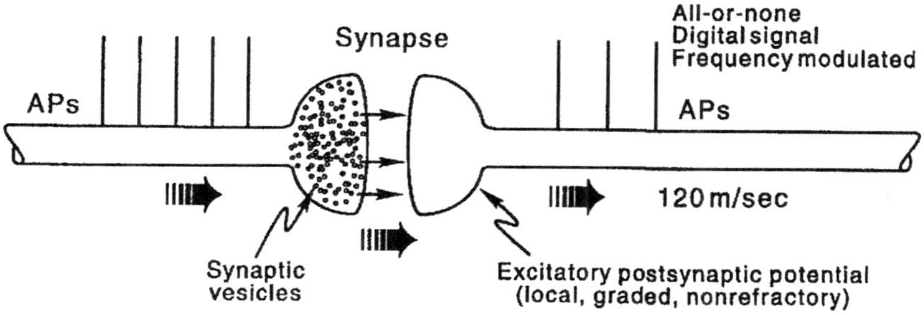

Figure XIII-2. Diagram of a chemical excitatory synapse between two nerve fibers. An action potential in the presynaptic fiber (left) brings about the release of the neurotransmitter at its nerve terminal. The transmitter molecules rapidly diffuse the short distance (e.g., 0.1 mm) across the synaptic cleft, bind to receptor sites on the postsynaptic membrane, and open the associated non-selective ion channels (Na^+-K^+ mixed conductance) complexed to the receptor (i.e., these are ligand-gated ion channels). The associated synaptic current depolarizes the postsynaptic membrane, producing the excitatory postsynaptic potential (EPSP), and this depolarization spreads passively into the adjacent conductile membrane (excitable) because of cable properties, thereby triggering one or more action potentials in the postsynaptic axon. The EPSPs are local, graded in amplitude, and non-refractory, whereas the action potentials are all-or-none (maximal), refractory, and propagated actively. Thus, the amplitude modulated (AM) synaptic process gives rise to a frequency modulated (FM) or digital signal. The strength of a biological response (e.g., contraction, secretion, or sensation) is a function of the frequency of the signal. That is, higher action potential frequency corresponds to stronger contraction or stronger sensation (in the sensory nervous system). (Modified from Fig. 2 in Sperelakis, N., 1993 (Cable Properties and Propagation. Ch 5, pp. 83-97; In Essentials of Physiology (N. Sperelakis & R.O. Banks, eds.), Little Brown and Co., Boston, 1993.)

(around the myelinated nerve fibers) causes propagation to become slowed and impaired in the affected nerve fibers, with an associated incoordination and partial paralysis.

Since propagating all-or-none APs are all very similar to each other (in shape, duration, amplitude, rate of rise, and propagation velocity), to make the signal stronger or weaker, the body increases or decreases the frequency of the APs, respectively. That is, the body uses a **frequency-modulated** (FM) system, rather than an amplitude-modulated (AM) system (Fig. XIII-2). It is a **digital system** comprised of yes/no identical signals. At each synapse, the signal becomes **graded** in amplitude rather than all-or-none: the greater the amplitude and duration of the local postsynaptic potential, the higher the frequency of APs triggered. The same is true of the local graded receptor potential generated at some sensory organs/receptors. Stronger signals translate into higher frequency of impulses, and weaker signals correspond to lower frequency of impulses.

B. Biological Fiber as a Cable

An **electrical cable** consists of two parallel conductors separated by insulation material, e.g., two copper wires separated by rubber. Usually one of the conductors is arranged as a tubular sleeve surrounding a central solid rod (wire), as depicted in Figure XIII-3A. The equivalent electrical circuit for a cable is shown in Figure XIII-3B: two parallel conductors (wires) separated by a transverse resistance (R), shown distributed along the length of the cable. The resistance of the conductors is so small compared to the transverse insulation resistance that they are assumed to be zero. In the case of the **biological cable** (long narrow nerve fiber or skeletal muscle fiber), one parallel conductor is the inside fluid (cytoplasm) and the second parallel conductor is the outside fluid

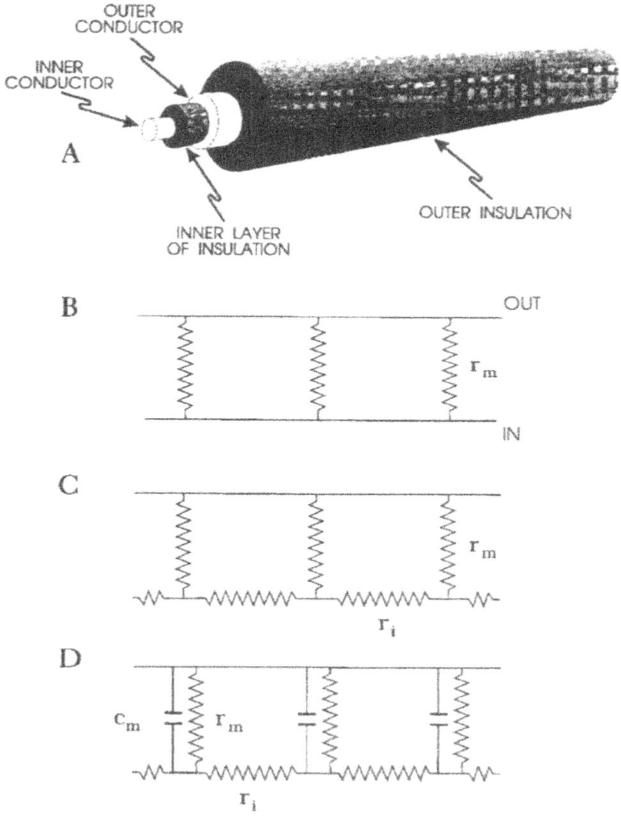

Figure XIII-3. A: A co-axial cable and associated electric equivalent circuit (B), the latter then being modified (C) to reflect the biologic cable (D). The co-axial cable consists of an inner conductor (e.g., copper wire) and an outer concentric conductor separated by a layer of insulation (e.g., rubber). **B:** The equivalent circuit for a co-axial cable. The inner and outer conductors depicted here have nearly zero resistances, and the transmembrane insulation resistance (R) is distributed along the length of the cable. **C:** In the biologic cable, the inner conductor is the cytoplasm (axoplasm or myoplasm), which is not of negligible resistivity, and therefore is depicted as R_i distributed along the length of the fiber. The outside longitudinal resistance is negligibly low, in comparison with the inside resistance, and therefore is drawn as zero resistance. The transverse insulation resistance is the cell membrane (R_m). **D:** Addition of the capacitance elements to the biologic cable, which arise due to the lipid bilayer matrix of the cell membrane (C_m).

surrounding (bathing) the cell (the interstitial fluid). In a submarine cable, the ocean can serve as the outer conductor. In the biological cable the tissue fluids, serving as the ocean, provide a fairly effective outside conductor. The inside conductor is not as effective, however. Because the conductivity of biological fluid is much less (i.e., much higher resistance) than that of copper wire, and because the cross-sectional area of the cell cytoplasm is so small, the inside longitudinal resistance is high and cannot be ignored (Fig. XIII-3C). The outside longitudinal resistance is relatively small, as compared to the inside (because of the larger volume (cross-sectional area)) of fluid available to carry the outside current, and therefore is assumed to be negligible.

In addition, there is a **stray capacitance** distributed along the length of the cable (Fig. XIII-3D), because a capacitance occurs when two parallel conductors ("plates") are separated by a high-resistance dielectric material. The **dielectric constant** of materials are related to vacuum, which is assigned a value of 1.0000; air has a value very close to vacuum. Oils have a dielectric constant of 3-6, and that of pure water is 81. The biological membrane, which has a matrix of phospholipid molecules, has a dielectric constant of about 5, typical of oils. The higher the dielectric constant, the higher the capacitance; the closer the parallel plates, the higher the capacitance. Since the biological membrane is so thin (approx. 70 Å or 7 nm, its capacitance is relatively high: all cell membranes have a membrane capacitance (C_m) of about 1.0 μ F/cm^2 (where F = Farad).

C. Length Constant

In the cable depicted in Figure XIII-3A-B, a voltage (or signal) applied at one end would be transmitted to a distant end with little or no decrement (diminution or attenuation), and the

length constant would be very long or nearly infinite. In the biological cable (Fig. XIII-3C-D), however, a signal applied at one end rapidly falls off (decays) in amplitude as a function of distance, with a relatively short length constant (λ). This decay in voltage is exponential (Fig. XIII-4A). An exponential gives a straight line on a semilogarthmic plot (log V vs. distance) (Fig. XIII-4B). In a cable, the relationship between the voltage at any distance (x) from the applied voltage (V_o) is:

$$V_x = V_o e^{-x/\lambda} \qquad \text{eq. 1}$$

Thus, when $x = \lambda$, $V_x = V_o\ 1/e^1 = V_o\ 1/2.718 = 0.37\ V_o$. Hence, the distance at which the voltage decays to 37% of the initial value gives the length constant. In nerve fibers and skeletal muscle fibers, λ has a value of only about 1 - 3 mm.

The mathematical solution to Equation 1 is:

$$V_x = \text{antilog}\left[\frac{2.303 \log V_o - \chi/\lambda}{2.303}\right] \qquad \text{eq. 1a}$$

Therefore, the relatively short λ, compared to the length of the neuron (e.g., 1.0 m for a lumbar anterior horn cell (motor neuron)) or skeletal muscle fiber means that a signal applied at one end (or mid-point) would fall off very quickly over distance along the fiber (Fig. XIII-5). If the length constant were 1.0 mm, then at 4 mm, the signal would become negligible. Hence, the electrical signal cannot be conducted passively in the biological cable, because it would decrement and disappear over relatively short distances. The AP (signal) is amplified to a constant value at each point (or each node) in the membrane. That is, conduction is active, not passive, with energy being put into the signal at each point to prevent any decay of the signal.

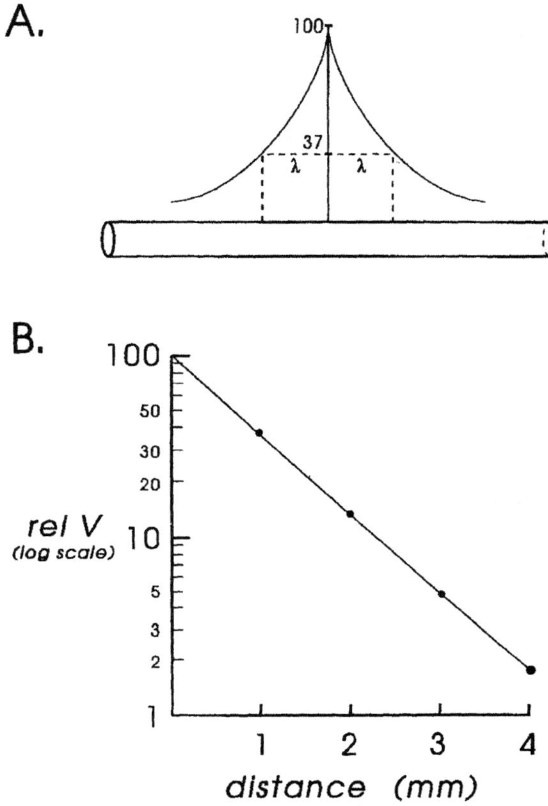

Figure XIII-4. Length constant (λ) of the biological cable (fiber). The exponential decay of voltage on both sides of an applied current (voltage) as a function of distance is diagrammed in part A. The distance at which the voltage falls to 1/e (or 36.7% of initial voltage (at $x = 0$)) gives the λ value. When the voltage is plotted on a log scale against distance, a straight line is obtained (part B). Redrawn from Figure 4 in Sperelakis, N., 1993 (Cable Properties and Propagation. Ch 5, pp. 83-97; In Essentials of Physiology (N. Sperelakis & R.O. Banks, eds.), Little Brown and Co., Boston, 1993.)

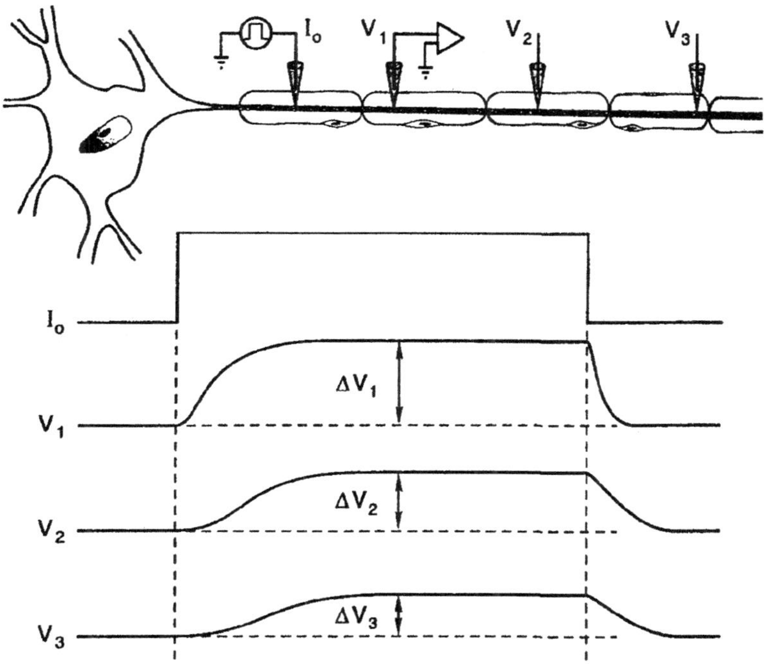

Figure XIII-5. Diagram of a motor axon and how the voltage signal would decay with distance if the neuron were only a passive cable (non-excitable). The voltage traces illustrate the voltage signals that would be simultaneously recorded at three different points along the axon (V_1, V_2, and V_3) from the site of injection of a rectangular current pulse (I_0). As depicted, the amplitude of the steady-state voltage pulse rapidly falls off with distance, because the length constant (λ) of the biological cable is short (e.g., 1 mm) compared to the length of the axon (e.g., 1,000 mm). Therefore, an active response of the cell membrane at each point is required to faithfully propagate the signal. The voltage recorded at point V_1 (ΔV_1) also illustrates that the membrane potential changes in an exponential manner, both at the beginning of an applied rectangular current pulse and at the end. This exponential charge results from the capacitance of the cell membrane, the time constant (τ) being a product of the resistance and capacitance ($R_m C_m$). The time it takes for the voltage to decay to $1/e$ (36.7%) at the end of the pulse, or to build up to 63.3% ($1 - 1/e$) at the beginning of the pulse, of the initial (maximal) value, gives the τ value.

The parameters of a cable that determine its length constant are the square root of the ratio of the transverse resistance to the sum of the inside and outside longitudinal resistances:

$$\lambda = \sqrt{\frac{r_m}{r_i + r_o}} = \sqrt{\frac{\Omega \cdot \mathrm{cm}}{\frac{\Omega}{\mathrm{cm}} + \frac{\Omega}{\mathrm{cm}}}} \qquad \text{eq. 2}$$

where r_m, r_i, and r_o are the normalized resistances for a unit length (1 cm) of fiber. For surface fibers of a nerve or muscle bundle bathed in a large volume conductor, r_o is negligibly small, and equation 2 reduces to:

$$\lambda = \sqrt{\frac{r_m}{r_i}} = \sqrt{\frac{R_m}{R_i}\frac{a}{2}} = \sqrt{\frac{\Omega \cdot \mathrm{cm}^2}{\Omega \cdot \mathrm{cm}}\mathrm{cm}} \qquad \text{eq. 3}$$

where a is the fiber radius and R_m and R_i are the membrane resistance and longitudinal cytoplasmic resistance normalized for both length (1 cm) and cell diameter. Thus, the greater the membrane resistance and the smaller the internal longitudinal resistance (larger cell diameter), the greater the λ value. Propagation velocity is a function of λ; e.g., larger diameter fibers propagate faster. Myelination increases the effective membrane resistance (R_m) and lowers the effective capacitance, and thereby increases propagation velocity.

D. Time Constant

Because of the large capacitance of the cell membrane (C_m), the membrane potential (E_m) cannot change instantaneously upon application of a step current pulse. Instead, E_m changes in an exponential (negative) manner (Fig. XIII-5), both on the charge and discharge. The membrane **time constant** (τ_m) is a product of the resistance (R_m) and capacitance of the membrane.

$$\tau_m = r_m c_m = R_m C_m \qquad \text{eq. 4}$$

$$\sec = \Omega \times \mathrm{F}$$

where Ω is the resistance in ohms, and F is the capacitance in Farads.

The **discharge** of the membrane (parallel RC network) is given by:

$$V_t = V_{\max} e^{-t/\tau} \qquad \text{eq. 5a}$$

where V_t is the voltage at any time t (at the site of current injection), and V_{max} is the final maximum voltage attained during the pulse. When $t = \tau$, $V_t = V_{max}\ 1/e^1 = V_{max}\ 1 / 2.718 = 0.37\ V_{max}$. Hence, the time at which the voltage decays to 37% of the initial (maximal) value gives the time constant, τ (Fig. XIII-5). In nerve fibers and skeletal muscle fibers, the time constant is about 1.0 msec.

When the membrane is **charging**, there **is** a similar exponential (negative) process, with the identical time constant (Fig. XIII-5). The corresponding relationship is given by:

$$V_t = V_{\max}(1 - e^{-t/\tau}) \qquad \text{eq. 5b}$$

The time it takes for build-up of the voltage to 63% (1 – 1/e = 1 – 0.37 = 0.63) of the final voltage gives the time constant. Thus, τ can be measured on the build-up (time to reach 63% of the final voltage) or on the decay (time to reach 37% of the initial voltage) of the pulse.

E. Local Potentials

In contrast to the active propagation of APs, synaptic potentials and sensory receptor potentials are not actively propagated. Such potentials decay exponentially (from their source of initiation) along the cell cable, as described above. Therefore, synaptic and receptor potentials are **local potentials**. When local potentials are in the depolarizing direction, they can give rise to APs which are propagated; when hyperpolarizing, they act to inhibit production of APs. These local potentials are similar to the **local excitatory response** in that both are confined to a local region, but the electrogenesis of the two are different. As stated above, the neuromuscular junction is an excitatory type of chemical synapse and produces excitatory postsynaptic potentials (EPSPs), known here as end-plate potentials (EPPs).

Most synaptic potentials are **graded**, i.e., they can add on one another, both in time and space (**temporal summation** and **spatial summation**), to produce larger responses. Larger synaptic potentials exert a greater stimulatory or inhibitory effect on the production of APs.

F. Local-Circuit Currents

This section examines the mechanism for the rapid propagation (conduction) of APs. Propagation occurs by means of the local-circuit currents that accompany the propagating APs, as depicted in Figure XIII-6. Such currents exist because, when two points are at a different potential (voltage) in a conducting medium, current (I) will flow between the two points, as governed by Ohm's law ($I = V/R$). At the peak of the AP in one region of the fiber, the inside of the membrane at that region becomes positive with respect to the outside. The inside is also positive with respect to the inside cytoplasm at a region downstream (from the active region). Therefore, current flows through the cytoplasm from the active region (**current source**) to the adjacent inactive region, then out of the fiber across the cell membrane, then through the interstitial fluid back to the active region (**current "sink"**), and finally through the membrane of the active region. This completes the closed loop for the current. The outward current through the membrane of the inactive region produces an IR voltage drop (Ohm's law), positive inside to negative outside, which acts to depolarize this region, because the polarity of the voltage drop is opposite to that of the resting potential (negative inside, positive outside). When the depolarization exceeds the threshold potential, an AP is triggered. Thus, the inactive region has been converted to an active region. This process is repeated in each segment of fiber, thus resulting in movement (propagation) of the impulse sequentially down the fiber.

With a propagating AP (in the middle region of a fiber; see Fig. XIII-6), there is also a small backflow of current internally, coupled with a corresponding small forward flow externally, associated with the repolarizing phase of the AP.

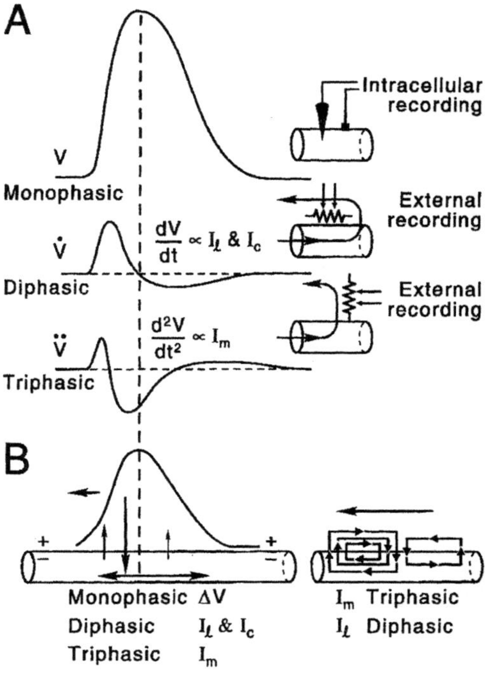

Figure XIII-6. Schematic representation of the first ($\dot{V}$) and second ($\ddot{V}$) time derivatives of the action potential spike and the longitudinal currents and radial currents associated with the propagating spike. *dV/dt* ($\dot{V}$) is proportional to the capacitive current ($I_c = C_m \cdot dV/dt$) and the longitudinal (axial) current, and is biphasic (A), having an intense forward phase and a less intense backward phase, as depicted in the diagram at the lower right (B). d^2V/dt^2 ($\ddot{V}$) is proportional to the radial transmembrane current (I_m), and is triphasic, having a moderately intense initial outward phase, then a very intense inward phase (the "current sink"), followed by a least intense second outward phase, as depicted in the lower diagrams (B). The arrows in the diagram at the lower left (B) also depict the three phases of the membrane current and the two phases of the axial current. *dV/dt* can be recorded externally by a pair of closely-spaced (relative to the spike wavelength) electrodes arranged parallel to the fiber axis, as illustrated. d^2V/dt^2 can be recorded by a pair of electrodes arranged perpendicular to the fiber axis, as depicted (A). The vertical dashed line indicates that when the slope of the spike goes to zero at the peak of the spike, *dV/dt* is zero; *dV/dt* is maximum at about the middle of the rising phase of the spike. Redrawn from Figure 6 in Sperelakis, N., 1993 (Cable Properties and Propagation. Ch 5, pp. 83-97; In Essentials of Physiology (N. Sperelakis & R.O. Banks, eds.), Little Brown and Co., Boston, 1993.)

Thus, as the AP propagates down the fiber, from right to left, there is a simultaneous double flow of local-circuit current: clockwise flow associated with the rising phase of the AP, and counter-clockwise flow associated with the repolarizing phase of the AP. The internal longitudinal current, sweeping past a transverse plane of the fiber, has two phases, first forward (right to left), and then reverse (left to right). The external longitudinal current also has two phases: first left to right and then right to left. The transverse membrane current, sweeping through a point in the membrane has three phases: first outward (still passive membrane), then inward (active membrane), and finally outward again (still active membrane).

It is the local-circuit current flow that enables us to record the electrocardiogram (ECG) from the surface of the body, and the electromyogram (EMG), electroencephalogram (EEG), and electroretinogram (ERG) from the body surface over the tissue of interest (skeletal muscle, brain, eye). The internal longitudinal current is confined to the cytoplasm of the fiber, but the external current can use whatever conducting fluid is available (e.g., the entire torso volume conductor), because of the principle of resistors in parallel giving a lower total resistance (or current takes the path of least resistance). Thus, this external local-circuit current causes the skin to be at different potentials, and these differences can be recorded (as the ECG, etc.).

G. Factors Affecting Propagation Velocity

The factors that determine active velocity of propagation (θ) include: (a) fiber diameter, (b) length constant (λ), (c) time constant (τ_m), (d) local-circuit current intensity, (e) threshold potential, and (f) temperature. Some of these factors are interrelated, such as fiber diameter and length constant (since λ

is proportional to square root of radius), and length constant with time constant (since both have a dependence on R_m). Propagation velocity is directly proportional to length constant and inversely proportional to time constant.

$$\theta \propto \frac{\lambda}{\tau_m} \qquad \text{eq. 6}$$

By substituting equation 3 for λ and equation 4 for τ_m, the following relationship is obtained:

$$\theta \propto \frac{\sqrt{a}}{C_m \sqrt{R_i}\ \sqrt{R_m}} \qquad \text{eq. 6a}$$

Thus, propagation velocity is directly proportional to the square root of fiber diameter or radius (a) and inversely proportional to membrane capacitance (C_m). The larger the fiber diameter, the lower is the absolute longitudinal resistance of the intracellular cytoplasm (law of resistors in parallel), and therefore the greater the amount of local-circuit current flowing longitudinally and the greater the length constant. For example, it is well known that the larger the diameter of nerve fibers, the faster they propagate. Equation 6a shows that if C_m can be reduced (e.g., by myelination), then θ should increase in proportion.

In addition, θ depends on the intensity of the local-circuit current, and hence on the rate of rise of the AP. The greater the AP rate of rise (max dV/dt), the greater is the longitudinal current and the transmembrane capacitive current (I_c). Therefore, all other factors being constant, faster rising APs propagate faster. The AP rate of rise depends on the density of the inward current-carrying fast Na^+ channels that are available, on C_m, and on temperature. max dV/dt decreases with increased C_m , with cooling, and with partial depolarization (due to the h_∞ vs. E_m relationship). Cooling slows the rate of all chemical reactions, especially those with a high Q_{10} (activation energy), such as the ion conductance changes in activated membrane.

Finally, the threshold potential (V_{th}) level affects propagation velocity. If the threshold were to be shifted to a more positive voltage (more depolarized), then it takes longer for a given point in the membrane to reach threshold (and explode) during propagation of an AP from upstream. A greater **critical depolarization** (difference between RP and V_{th}) is required to bring the membrane to threshold. Therefore, propagation velocity would be slowed.

As stated above, some of these factors are inter-related, and some actually exert opposing effects.

The discussion presented above applies to non-myelinated nerve axons and skeletal muscle fibers. In myelinated nerve fibers, propagation velocity is greatly increased by the myelin sheath, as discussed in the following section.

H. Saltatory Conduction

The nerve cable has been vastly improved by the evolutionary development in vertebrates of **myelination**. The **myelin sheath** improves the cable by increasing the effective R_m by about a hundred-fold and decreasing the effective C_m by about a hundred-fold. This increases the length constant, λ, and tends to decrease the time constant (τ_m). Since τ_m tends to increase with the increase in effective R_m (due to the myelin sheath), the decrease in C_m counteracts this effect, thus acting to hold τ_m almost constant. Thus, equations 6 and 6a predict that propagation velocity should increase with myelination.

Thus, one consequence of myelination is that **propagation velocity** is greatly increased. A second consequence is that the **energy cost** of signalling is greatly decreased. The energy cost is reduced because passive ion leaks are limited, and active current losses are restricted to the small nodes of Ranvier which are spaced relatively far apart: 0.5 to 2.0 mm internodal length. At each node, the length of exposed (naked) cell membrane is only a few microns. Therefore, the degree of energy-requiring active ion transport (Na-K and Ca^{2+}), required to maintain the steady-state ion distributions and to hold the system in a state of high potential energy, is greatly reduced. For example, the amount of Na^+ gained and K^+ lost per impuse is reduced as a result of myelination. The amount of oxidative metabolism in myelinated fibers reflects this lowered energy requirment.

The myelin sheath is produced by the wrapping of the **Schwann cell** repeatedly around the nerve fiber in a spiral, forming 50–300 wrappings. For purpose of our discussion, we will assume an average of 100 wrappings. The myelin sheath covers the nerve axon as a coat sleeve, and is interrupted at each node. The cytoplasm of the Schwann cell is nearly completely extruded during formation of the myelin sheath, so that the sheath consists essentially of 100 cell membranes in series. Because of the law of resistors in series, the effective transmembrane resistance is increased 100-fold. Because of the law of capacitors in series, the effective capacitance is reduced 100-fold. Since λ is directly proportional to the $\sqrt{R_m}$, λ is increased accordingly. As described above, increasing λ and lowering C_m increase θ.

However, the myelin sheath is interrupted at every node of Ranvier. That is, the axon is bare (nude) at each node. The internodal distance is about 0.5–2.0 mm (depending on fiber diameter), and the width (length) of each node is only about 0.5–3 μm. The node is an annulus around the entire perimeter of the fiber.

Myelinated nerves usually have an optimal amount of myelin, which is such that the ratio of diameter of axis cylinder (naked axon) to total fiber (including myelin sheath) is about 0.6–0.7. Assuming a maximal total diameter feasible (the body must pack many circuits within a limited space (e.g., sciatic nerve bundle)), a greater fraction of myelin, by infringing on the diameter of the axis cylinder, would raise r_i too high, causing θ to diminish. Thus, there are two opposing factors in determining the degree of optimal thickness of the myelin sheath: (a) the more the myelin, the greater the increase in λ and decrease in C_m; (b) the more the myelin, the lesser the diameter of the axis cylinder and therefore the higher the r_i.

In **saltatory conduction** (Lat. saltere = to jump), the impulse jumps from one node to the next. The internodal membrane does not fire an AP. This is due to two reasons: (a) the internodal membrane is much less excitable (e.g., much fewer or no fast Na^+ channels), and (b) the depolarization of the neuron cell membrane at the internodal region is only about 1/100th of that at the node. The latter occurs because the IR voltage drop across the internodal cell membrane is only 1/100th of that across the entire series resistance network (neuron cell membrane plus many layers of Schwann cell membrane) because of Kirchhoff's laws dealing with voltage drops across resistors in series. Even though the internal potential in the internodal region swings positive (e.g., to +30 mV) when the adjacent nodes fire, the potential at the outer surface of the neuronal membrane also swings nearly as positive (e.g., to +29 mV). Therefore, the depolarization of the neuronal membrane at the internode is only about 1 mV, which is well below threshold, and so the internodal membrane does not fire. The potential that controls the membrane conductances (activates the voltage-dependent ion channels) is the p.d. (potential difference) directly across the membrane and not the absolute potential on either side.

As stated above, the internodal membrane has only few or no fast Na^+ channels. Since the cell membrane is fluid and proteins can diffuse (float) laterally in the lipid bilayer matrix, the fast Na^+ channel proteins are confined (at high density) in the nodal region by special anchoring proteins (e.g., anchorin). These proteins tether the ion channel proteins to the cytoskeletal framework, thus preventing their lateral movement into the internodal membrane.

The effect of myelin and saltatory propagation is to make propagation much faster. For example, a 20 μm diameter myelinated nerve fiber conducts even faster than a 1,000 μm (1.0 mm) diameter nonmyelinated nerve fiber (e.g., the giant axon in squid, lobster, earthworm): 120 m/sec vs. ca. 25–50 m/sec. Thus, invertebrates, in order to achieve fast conduction in some essential circuits, have to resort to giant neurons, resulting in a lower r_i and hence fast conduction. Because of space/size limitations, only a few critical neurons can be made giant diameter. In vertebrates, on the other hand, a large fraction of the nerve fibers in our peripheral nerves are myelinated for fast propagation.

As was described previously, in nonmyelinated axons and skeletal muscle fibers, θ should vary with the square root of the cell diameter or radius ($a^{0.5}$). In myelinated axons, propagation velocity varies with the first power of the cell radius (a^1), because propagation velocity (θ) varies with the length constant squared:

$$\theta \propto \frac{\lambda^2}{\tau_m} \propto \frac{a}{2R_i C_m} \qquad \text{eq. 7}$$

The dependence of conduction velocity on diameter of myelinated and non-myelinated fibers is summarized in Table XIII-1.

I. Wavelength of the Impulse

We can calculate the **wavelength** of the action potential, which is the length of the axon simultaneously undergoing some portion of the action potential. The wavelength is equal to propagation velocity (θ) times the duration of the action potential (APD_{100}):

$$\text{Wavelength} = \theta \times APD_{100} \qquad \text{eq. 8}$$

$$\text{cm} = \frac{\text{cm}}{\text{sec}} \times \text{sec}$$

$$\text{distance} = \text{velocity} \times \text{time}$$

Note the similarity of this relationship given in equation 8 to that for the wavelength of an electromagnetic radiation:

$$\text{wavelength} = \frac{\text{velocity of light}}{\text{frequency}} \qquad \text{eq. 8a}$$

wavelength = velocity of light/frequency of the radiation. The reciprocal of frequency is the period (duration of one cycle).

The wavelength in a large myelinated nerve axon is about 12 cm: 120 m/sec x 1.0 msec. In a skeletal muscle fiber, it is about 1.8 cm (6 m/sec x 3.0 msec). In a fine smooth muscle bundle, the wavelength is only about 1.5 mm (5 cm/sec x 30 msec).

J. External Recording of Action Potentials

As discussed above, local-circuit currents accompany the propagating AP in each fiber. The intracellular and extracellular longitudinal currents are **biphasic**: initially in the forward direction intracellularly and followed by the reverse direction. The forward direction current is intense (high current density) and the reverse direction current is weak (low current density). The transmembrane radial currents are **triphasic**: the first phase is outward (moderate intensity), second phase is inward (high intensity), and the third phase is outward (low intensity). The first phase (outward) gives rise to the passive exponential foot of the AP, and is due to the passive cable spread of voltage and current. The second phase (inward) corresponds to the large inward fast Na^+ current, which occurs during the later portion of the rising phase and peak of the action potential. The third phase (outward) corresponds to the net outward current (K^+), which occurs during the repolarizing phase.

These longitudinal and radial currents can be recorded by suitably-placed external electrodes. The extracellular **longitudinal currents** can be recorded by two electrodes (bipolar) placed close together along the length of the fiber. If the inter-electrode distance is short (relative to the wavelength), an approximate first (time) derivative of the action potential is obtained (Fig. XIII-6).

The extracellular **radial currents** can be recorded by two electrodes placed close together in a plane perpendicular to the fiber axis. This gives an approximation of the second (time) derivative of the AP (Fig. XIII-6).

The internal axial currents are confined to the cytoplasm, whereas the external longitudinal currents can use the entire interstitial fluid space of the nerve bundle or muscle and even the entire torso (so-called "volume conductor"), since current takes the path of least resistance (law of resistors in parallel). This fact allows the recording of the ECG from the body surface and the EMG from the skin overlaying an activated skeletal muscle. The ECG and EMG consist essentially of diphasic potentials, reflecting the external longitudinal currents during propagation of action potentials.

For **biphasic recording** in which the two phases are about equal, the two external electrodes can be placed far apart (with respect to the wavelength) along a nerve or muscle fiber. The proximal electrode records the wave of negativity (associated with the propagating action potential) first, and then returns to isopotential. When the wave reaches the second electrode, the wave is recorded by it in reversed polarity (because current flow through the voltmeter is reversed).

Now if the action potential is prevented from reaching the second (distal) electrode, by crushing this region of the fiber or elevating $[K]_o$ to depolarize it, then a **monophasic recording** is obtained (Fig. XIII-7). This recording most resembles the true action potential which is recorded by a microelectrode impaled into a fiber to record the transmembrane potential (Fig. XIII-7), but, of course, is much smaller in amplitude.

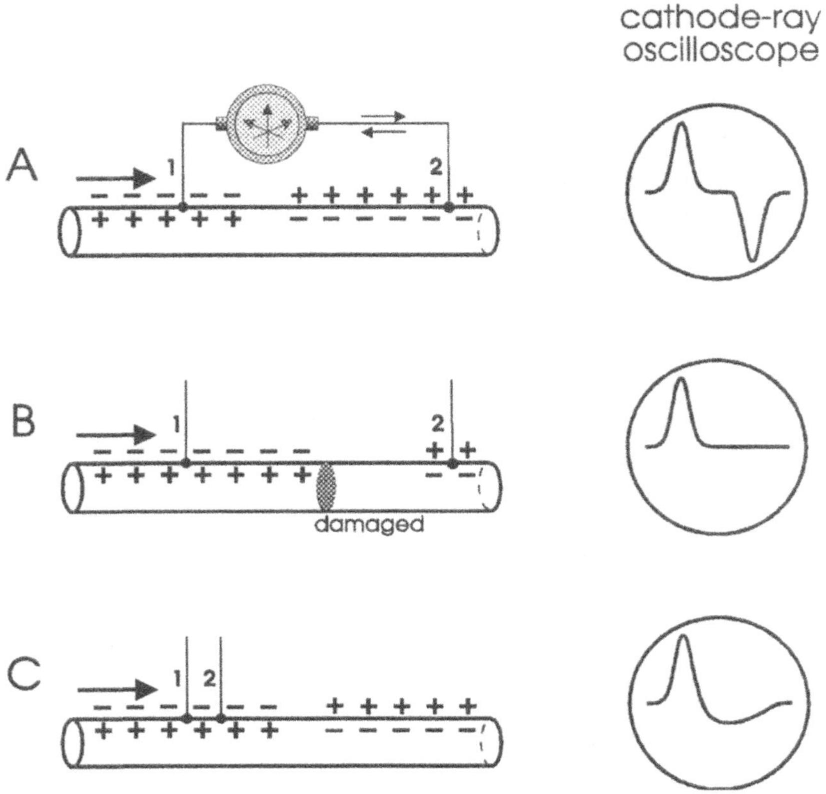

Figure XIII-7. Diagram of the waveforms that would be recorded externally during propagation of an action potential in a single fiber. In part A, the two electrodes are far apart (relative to the action potential wavelength), and so a biphasic recording is obtained, with the two phases being symmetrical and separated by an isopotential segment. The two phases are due to the current flow through the voltmeter recorder being first in one direction and then in the opposite direction. If the fiber between these two electrodes is damaged (e.g., by crushing) or depolarized (by elevated $[K]_0$), so the action potential can't sweep past the second electrode (#2), then this second phase is prevented and the recording is monophasic, as depicted in part B. If the two electrodes are brought progressively closer, then the isopotential segment depicted in part A would shorten and disappear. As shown in part C, if electrode #2 is brought very close to electrode #1, so that the inter-electrode distance is short relative to the wavelength, then the second phase is smaller than the first phase, and the record resembles the first derivative of the true action potential.

K. Compound Action Potential

When one records the action potentials externally, the records are **graded** and not all-or-none, as in the case of the true action potentials recorded intracellularly from single fibers. That is, the recorded "action potential" gets larger and larger, up to a maximum amplitude, as the intensity of stimulation is increased. This is the so-called **compound action potential**. It is graded because, as a greater and greater fraction of the fibers are activated, the external longitudinal currents associated with the all-or-none action potential in each fiber cut across the recording electrodes, and thereby produce a larger signal. The **action potential in each individual fiber is always all-or-none**. The amplitude of the signal is determined by the resistance between the electrodes multiplied by the amount of current flowing through this resistance ($V = IR$). The recording of compound action potentials is diagrammed in Figure XIII-8.

The compound action potentials can be demonstrated by recording the EMG from a human subject if one electrode is placed on the skin of the ventral forearm, and the other (reference) electrode on the wrist of the same arm, then as the subject voluntarily produces stronger and stronger contractions to flex the hand, the electrical signals picked up get greater and greater in amplitude and frequency. The amplitude gets larger because more muscle fibers are simultaneously activated. This is known as **fiber recruitment**. The frequency increases because the motor nerves fire at a higher frequency, causing the muscle fibers to fire at a higher frequency, to produce a more powerful complete tetanic contraction.

L. Summary

The biological cable (i.e., nerve fiber or skeletal muscle fiber) is a relatively poor cable with a short length constant, although it is the best possible given the biological circumstances. Therefore, for faithful and rapid signal transmission over long distances, energy must be put into the system at each point along the way. The system evolved is that of action potential generation, which are all-or-none signals of constant amplitude and constant propagation velocity, and having refractory periods and sharp thresholds. This is a frequency-modulated system, in which increasing strength of sensation or motor response follows from an increase in frequency of the action potential signals.

Action potential propagation occurs by means of the local-circuit currents. The transmembrane current has three phases: outward, inward, and outward; the internal and external longitudinal currents have two phases: forward and backward (for internal) or backward and forward (for external). The external currents use the path of least resistance, enabling electrograms (e.g., ECG, EMG) to be recorded from the body surface. The compound action potential is graded in amplitude, reflecting the summation of the external currents generated from each fiber that is activated: the more fibers simultaneously activated, the greater the amplitude of the electrogram signal.

Propagation velocity is faster the larger the diameter of the fiber, the longer the length constant, and the lower its time constant and capacitance. The myelin sheath evolved by vertebrates enables much faster propagation velocity and at a lower energy cost. Myelination raises the effective membrane resistance and lowers the effective capacitance, and excitability occurs only at the short nodes of Ranvier that periodically interrupt the myelin sheath. Therefore, the action potential signal jumps from node to node in a saltatory pattern of conduction.

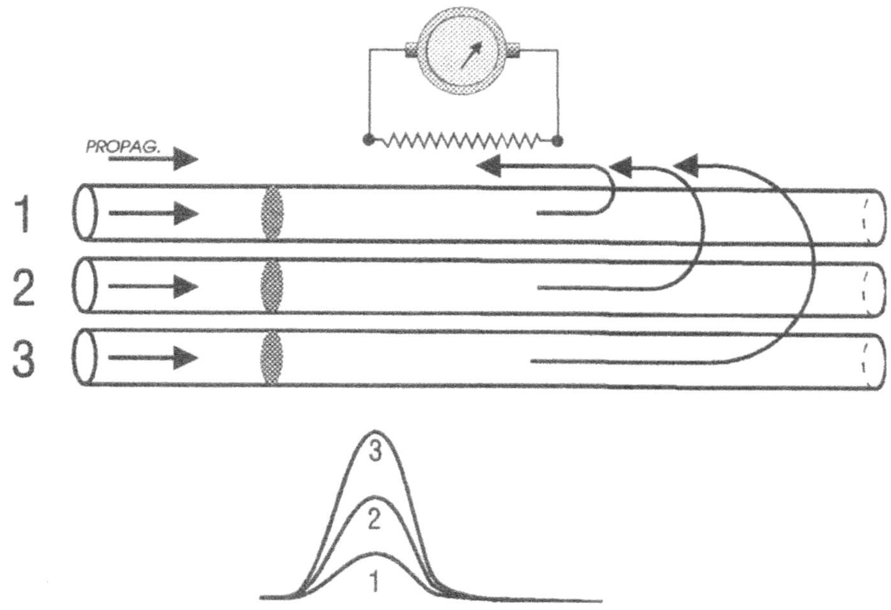

Figure XIII-8. Diagram of a compound action potential in an isolated nerve trunk, such as the frog sciatic nerve, externally recorded by a pair of longitudinal electrodes. The voltmeter (oscilloscope) recorder records the voltage (*IR*) drop across the resistance (fluid) between the two electrodes. If only fiber #1 is activated, the current passing between the electrodes is small, and the voltage recorded is small. If fiber #2 is simultaneously activated with fiber #1, then the amount of current is doubled, and the voltage is doubled. When all three fibers are simultaneously activated, the current is tripled, and the voltage is tripled. Therefore, the externally-recorded compound action potential is graded because it reflects the electrical activity of numerous fibers, each of which produces an all-or-none (non-graded) action potential.

Chapter XIV

SUMMARY

In this book, I have attempted to discuss most of the factors that determine or influence the resting potential of cells. The structural and chemical composition of the cell membrane was briefly examined and correlated with the membrane's resistive and capacitative properties. I have discussed and summarized most of the factors that determine the intracellular ion concentrations in myocardial cells under steady-state conditions. These factors include the Na-K coupled pump, the Ca/Na exchange reaction, a possible sarcolemmal Ca pump, and the intracellular polyvalent macroelectrolytes. The Na-K pump is mediated by the membrane-bound (Na-K)-ATPase enzyme. This enzyme requires both Na^+ and K^+ for activity, and transports 3 Na^+ ions outward and usually 2 K^+ ions inward per ATP hydrolyzed. Cardiac glycosides are specific blockers of this transport ATPase. The Na-K pump is not directly related to excitability, but only indirectly related by its role in maintaining the Na^+ and K^+ concentration gradients (i.e., blockage of the pump does not produce an immediate large depolarization and cessation of action potentials and of contractions). The Ca/Na exchange reaction seems to be carrier mediated and may be driven by the Na^+ electrochemical gradient; i.e., in the long run the energy for pumping out internal Ca^{2+} by this mechanism comes from the (Na-K)-ATPase. The Ca/Na exchange reaction exchanges one internal Ca^{2+} ion for two or three external Na^+ ions.

The mechanism whereby the ionic distributions give rise to diffusion potentials was discussed, as were the factors that determine the magnitude and polarity of each ionic equilibrium potential. The equilibrium potential for any ion and the transmembrane potential determine the total electrochemical driving force for that ion, and the product of this driving force and membrane conductance for that ion determine the net ionic current or flux. The net ionic movement can be inward or outward across the membrane, depending on the direction of the electrochemical gradient.

The key factor that determines the resting potential — in the absence of any electrogenic pump potential contributions and for fixed ionic distributions — is the relative permeability of the various ions, particularly of K^+ and Na^+, i.e., the P_{Na}/P_K ratio (or g_{Na}/g_K ratio), as calculated from the Goldman constant field equation. The major physiological ions that have some effect on the resting potential or on the action potentials are K^+, Na^+, Ca^{2+}, and Cl^-. The Ca^{2+} electrochemical gradient has only a small direct effect on the resting potential, although low external Ca^{2+} can affect the permeabilities and conductances for the other ions, such as Na^+ and K^+. Elevation of internal Ca^{2+} can sometimes increase the permeability to K^+. The evidence suggests that the Cl^- ion is passively distributed according to the membrane potential, i.e., not actively transported in skeletal muscle; in cardiac muscle also it is usually assumed that Cl^- is passively distributed. Such a passive distribution is not clear at the present time, however, because there is some evidence indicating that $[Cl^-]_i$ is about 2 or 3 times as great as that predicted according to the E_m , thus giving an E_{Cl} value of about -35 to -50 mV compared to a resting potential of about -80 mV. Before one can conclude that there is a Cl^- pump directed inward, however, the calculated E_{Cl} (concentrations corrected for activity coefficients) must be proved to be significantly more positive than the mean resting potential of the cell averaged over a period of time; i.e., for example, any spontaneous action potentials must be taken into account. if Cl^- is passively distributed, it cannot determine the resting potential, but transient net movements of Cl^- ions, e.g., during the action potential, can and do affect the E_m , particularly when g_{Cl} is high.

The elevation of $[K]_o$ to more than the normal concentration of about 3–4 mM decreases the K^+ equilibrium potential (E_K) as predicted from the Nernst equation ($[K]_i$ about constant), and depolarization is produced. Sometimes, however, some hyperpolarization is first produced at a $[K]_o$ level of about 5–10 mM; in addition, lowering $[K]_o$ to values of 1.0–0.1 mM often produces a prominent depolarization. These effects are usually explained on the basis that: (a) g_K is a function of P_K and $[K]_o$, (b) P_K is lowered in low $[K]_o$ and elevated in higher $[K]_o$, and (c) an electrogenic Na-K pump potential is inhibited at a low $[K]_o$ and stimulated at a higher $[K]_o$.

The resting potential not only is the potential energy storehouse that is drawn upon for propagation of the action potentials but because the membrane voltage-dependent cationic channels are inactivated with sustained depolarization, the rate of rise of the action potential, and hence propagation velocity, are critically dependent on the level of the resting potential. For example, a small elevation of K^+ concentration in the blood has dire consequences for functioning of the heart.

The Na-K pump is electrogenic, depending on the coupling ratio of Na^+ pumped out to K^+ pumped in. The addition to the resting potential depends on the magnitude of the membrane resistance. The electrogenic pump potential may be in parallel to the net ionic diffusion potential (E_{diff} , determined by the ionic equilibrium potentials and by the relative permeabilities). The contribution of the electrogenic pump potential to the measured resting potential of myocardial cells is generally small (only a few millivolts) so that the immediate depolarization produced by complete Na-K pump stoppage with cardiac glycosides is only a few millivolts. Of course, long-term pump stoppage produces a larger and larger depolarization as the ionic gradients are dissipated. The rate of Na-K pumping, and hence the magnitude of the electrogenic pump contribution to E_m, is controlled by $[Na]_i$ and by $[K]_o$, among other factors. The electrogenic pump potential might be physiologically important to the heart under certain conditions that tend to depolarize the cells, such as transient ischemia or hypoxia; in such cases the actual depolarization produced may be less because of a constant pump potential in parallel with a diminishing E_{diff}. The electrogenic pump potential may also affect automaticity of the nodal cells.

During embryonic development of the heart, dramatic changes in membrane electrical properties occur. The cells in the young tubular hearts have properties much like smooth muscle cells, namely, a low P_K that causes a high P_{Na}/P_K ratio and that therefore causes a low resting potential and a high degree of automaticity, and an absence of fast Na^+ channels. The inward current during the action potential is carried through slow cationic channels that are insensitive to tetrodotoxin. The P_K increases gradually during the first few days, and functional fast Na^+ channels appear suddenly during development and increase in number slowly thereafter until near the end of gestation.

REFERENCES

Abe M, Yamazaki N, Suzuki Y, Kobayashi A, Ohta H. 1984. Effect of palmitoyl carnitine on Na^+, K^+-ATPase and adenylate cyclase activity of canine myocardial sarcolemma. J Mol Cell Cardiol 16:239-245.

Adams RJ, Cohen DW, Gupte S, Johnson JD, Wallick ET, Wang T, Schwartz A. 1979. In vitro effects of palmitylcarnitine on cardiac plasma membrane Na,K-ATPase, and sarcoplasmic reticulum Ca^{2+}-ATPase and Ca^{2+} transport. J Biol Chem 254:12404-12410.

Adrian RH, Bryant SH. 1974. On the repetitive discharge in myotonic muscle fibers. J Physiol London 240:505-515.

Adrian RH, Freygang WH. 1962. The potassium and chloride conductance of frog muscle membrane. J Physiol London 163:61-103.

Adrian RH, Slayman CL. 1966. Membrane potential and conductance during transport of sodium, potassium and rubidium in frog muscle. J Physiol London 184:970-1014.

Akiyama T, Fozzard HA. 1975. Influence of potassium ions and osmolality on the resting membrane potential of rabbit ventricular papillary muscle with estimation of the activity and the activity coefficient of internal potassium. Circulation Res 37:621-629.

Alanis J, Benitez D. 1967. The decrease in the automatism of the Purkinje pacemaker fibers provoked by high frequencies of stimulation. Japan J Physiol 17:556-571.

Andersen NB, Gravenstein JS. 1965. Effects of local anesthetics on sodium and potassium in human red cells. J Pharmacol Exptl Therap 147:40-47.

Appel, SH, Roses AD. 1976. Membrane biochemical studies in myotonic muscular dystrophy. In Membranes and Disease, Bolis L, Hoffman JF, Leaf A,eds. New York: Raven Press, pp 183-195.

Arlock P, Katzung BG. 1985. Effects of sodium substitutes on transient inward current and tension in guinea-pig and ferret papillary muscle. J Physiol London 360:105-120.

Armstrong CM, Binstock L. 1965. Anomalous rectification in the squid giant axon injected with tetraethylammonium ion chloride. J Gen Physiol 48:859-872.

Bahinski A, Nakao M, Gadsby DC. 1988. Potassium translocation by the Na/K pump is voltage-insenstive. Proc Natl Acad Sci USA 85:3412-3416.

Baker PF. 1964. An efflux of ninhydrin-positive material associated with the operation of the Na^+ pump in intact crab nerve in Na^+-free solution. Biochim Biophys Acta 88:458- 460.

Baker PF, Glitsch HG. 1973. Does metabolic energy participate directly in the Na^+-dependent extrusion of Ca^{2+}-Ca^{2+} ions from squid giant axons? J Physiol London 233:44P-46P.

Baker PF, Blaustein MP, Hodgkin AL, Steinhardt RA. 1969a. Influence of calcium on sodium efflux in squid axons. J Physiol London 200:431-458.

Baker PF, Blaustein MP, Keynes RD, Manil J, Shaw TI, Steinhardt RA. 1969b. The ouabain-sensitive fluxes of sodium and potassium in squid giant axon. J Physiol London 200:459-496.

Bangham AD, Standish MM, Watkins JC. 1965. The action of steroids and streptolysin S on the permeability of phospholipid structures to cations. J Mol Biol 13:253-259.

Barnett RE, Palazzotto J. 1974. Mechanism of the effects of lipid phase transitions on the Na, K-ATPase, and the role of protein conformational changes. Ann NY Acad Sci 242:69-76.

Barr L, Dewey MM, Berger W. 1965. Propagation of action potentials and the structure of the nexus in cardiac muscle. J Gen Physiol 48:797-823.

Belardinelli L, Harder DR, Sperelakis N, Rubio R, Berne RM. 1979. Cardiac glycoside stimulation of inward Ca^{2+} current in vascular smooth muscle of canine coronary artery. J Pharm Exptl Therap 209:62-66.

Benndorf K, Friedrich M, Hirche H. 1991. Reoxygenation-induced arrhythmogenic transient inward currents in isolated cells of the guinea-pig heart. Pflugers Arch 418:248-260.

Berlin JR, Cannell MB, Lederer WJ. 1989. Cellular origins of the transient inward current in cardiac myocytes. Circ Res 65:115-126.

Bernard C. 1975. Establishment of ionic permeabilities of the myocardial membrane during embryonic development of the rat. In Developmental and Physiological Correlates of Cardiac Muscle, Lieberman M, Sano T, eds. New York: Raven Press, pp 169-184 (Perspectives in Cardiovascular Research, vol. 1).

Blaustein MP. 1976a. The ins and outs of calcium transport in squid axons: internal and external ion activation of calcium efflux. Federation Proc 35:2574-2578.

Blaustein MP. 1976b. Sodium-calcium exchange and the regulation of cell calcium in muscle fibers. Physiologist 19:525-540.

Blaustein MP, Hodgkin AL. 1969. Effect of cyanide on the efflux of calcium from squid axons. J Physiol London 200:497-527.

Boardman L, Hume SP, Lamb JF, McCall D, Newton JP, Polson JM. 1975. Genetic control of sodium pump density. In Developmental and Physiological Correlates of Cardiac Muscle. Lieberman M, Sano T, eds. New York: Raven Press, 1975, pp 127-138 (Perspectives in Cardiovascular Research, vol. 1).

Bockris JO'M, Conway BE. 1956. Modern aspects of electrochemistry. London: Butterworth & Co. (No. 3).

Bryant SH, Morales-Aguilera A. 1971. Chloride conductance in normal and myotonic muscle fibers and the action of monocarboxylic acids. J Physiol London 219:367-383.
Busselen P, Carmeliet E. 1973. Protagonistic effects of Na and Ca on tension development in cardiac muscle at low extracellular Na concentration. Nature New Biol 243:57-58.
Carmeliet EE. 1961. Chloride ions and the membrane potential of Purkinje fibers. J Physiol London 156: 375-388.
Carmeliet E, Bosteels S. 1969. Coupling between Cl flux and Na or K flux in cardiac Purkinje fibers: influence of pH. Arch Intern Physiol Biochim 77:57-72.
Carmeliet E, Janse M. 1965. Intracellular chloride concentration in cat papillary muscles: influence of external K concentration. Arch Intern Physiol Biochim 73:174-175.
Carmeliet E, Verdonck F. 1977. Reduction of potassium permeability by chloride substitution in cardiac cells. J Physiol London 265:193-206.
Carmeliet EE, Horres CR, Lieberman M, Vereecke JS. 1976. Developmental aspects of potassium flux and permeability of the embryonic chick heart. J Physiol London 254:673-692.
Charnock JS, Cook DA, Almeida AF, To R. 1973. Activation energy and phospholipid requirements of membrane-bound adenosine triphosphatase. Arch Biochem Biophys 159:393-399.
Chesnais JM, Coraboeuf E, Sauviat MP, Vassas JM. 1975. Sensitivity to H, Li, and Mg ions on the slow inward sodium current in frog atrial fibers. J Mol Cellular Cardiol 7:627-642.
Christoffersen GRJ. 1972. Steady state contribution of the Na, K-pump to the membrane potential in identified neurons of a terrestrial snail, Helix pomatia. Acta Physiol Scand 86:498-514.

Cole KS. 1968. Membranes, Ions and Impulses: A Chapter of Classical Biophysics. Berkeley: Univ. of California Press.

Cole KS, Curtis HJ. 1939. Electric impedance of the squid giant axon during activity. J Gen Physiol 22:649-687.

Colquhoun D, Neher E, Reuter H, Stevens CF. 1981. Inward current channels activated by intracellular Ca in cultured cardiac cells. Nature London 294:752-754.

Connelly CM. 1959. Recovery processes and metabolism of nerve. Rev Mod Phys 31:475-484.

Coraboeuf E, Deroubaix E, Hoerter J. 1976. Control of ionic permeabilities in normal and ischemic hearts. Circulation Res8:(Suppl. 1)192-198.

Cornelius F. 1991. Functional reconstitution of the sodium pump. Kinetics of exchange reactions performed by reconstituted Na/K-ATPase. Biochim Biophys Acta 1071:19-66.

Corr PB, Snyder DW, Cain ME, Drafford WA, Gross RW, Sobel BE. 1981. Electrophysiological effects of amphiphiles on canine Purkinje fibers. Implications for dysrhythmias secondary to ischemia. Circ Res 49:354-363.

Cranefield PF. 1975. The Conduction of the Cardiac Impulse. Mt Kisco, NY: Futura.

Cross SB, Keynes RD, Rybova R. 1965. The coupling of sodium efflux and potassium influx in frog muscles. J Physiol London 181:865-880.

Danielli DF. 1943. The theory of penetration of a thin membrane. Appendix In The Permeability of Natural Membranes, Davson H, Danielli DF, eds. London: Cambridge Univ. Press.

Davson H. 1964. Textbook of General Physiology. Boston: Little, Brown.

DeBeyer JM, DeMan JCH, Persijn JP. 1962. ATP-ase activity on the intercalated disc and C_Z bands of mouse heart muscle. J Cell Biol 13:451-456.

Degani C, Boyer PD. 1973. A borohydride reduction model for characterization of the acyl phosphate linkage in proteins and its application to sarcoplasmic reticulum adenosine triphosphatase. J Biol Chem 248:8222-8226.

DeHaan RL, Gottlieb SH. 1968. The electrical activity of embryonic chick heart cells isolated in tissue culture singly or in interconnected cell sheets. J Gen Physiol 52:643-665.

DeMello WC. 1963. Role of chloride ions in cardiac action and pacemaker potentials. Am J Physiol 205:567-575.

DeMello WC, Motta GE, Chapeau M. 1969. A study of the healing-over of myocardial cells of toads. Circulation Res 24:475-487.

Deweer P. 1970. Effects of intracellular ADP and P_i on the sodium pump of squid giant axon. Nature 226:1251-1252.

DeWeer P, Gadsby DC, Rakowski RF. 1988. Voltage-dependence of the Na/K pump. Annu Rev Physiol 50:225-241.

Dipolo R. 1973. Calcium efflux from internally dialyzed squid giant axons. J Gen Physiol 62:575-589.

Dreifuss JJ, Girardier L, Forssmann WG. 1966. Etude de la propagation de l'excitation dans le ventricule de rat au moyen de solutions hypertoniques. Arch Ges Physiol 292:13-33.

Droogmans G, Casteels R. 1976. Membrane potential and ion transport in smooth muscle cells. In Physiology of Smooth Muscle, Bulbring E, Shuba MF, eds. New York: Raven, pp 11-18.

Dudel J, Peper K, Rutdel R, Trautwein W. 1967. The dynamic chloride component of membrane current in Purkinje fibers. Pflugers Arch Ges Physiol 295:197-212.

Ehara T, Matsuoka S, Noma A. 1989. Measurement of reversal potential of Na^+-Ca^{2+} exchange current in single guinea-pig ventricular cells. J Physiol London 410:227-249.

Eisner DA, Lederer WJ. 1980. Characterization of the electrogenic sodium pump in cardiac Purkinje fibres. J Physiol 303:441-474.

Eisner DA, Lederer WJ. 1985. Na-Ca exchange: Stoichiometry and electrogenicity. Am J Physiol 248(Cell Physiol 17):C189-C202.

El-Sharkawy TY, Daniel EE. 1975. Electrogenic sodium pumping in rabbit small intestinal smooth muscle. Am J Physiol 229:1277-1286.

Fabiato A, Fabiato F. 1978. Effects of pH on the myofilaments and the sarcoplasmic reticulum of skinned cells from cardiac and skeletal muscles. J Physiol London 276:233-255.

Fedida D, Noble D, Rankin AC, Spindler AJ. 1987. The arrhythomogenic transient inward current I_{Ti} and related contraction in isolated guinea-pig ventricular myocytes. J Physiol London 392:523-542.

Ferrier GR. 1991. Digitalis toxicity. In Basic and clinical electrophysiology and pharmacology of the heart, Dangman KH, Miura DS, eds. New York: Marcel Dekker.

Forbes MS, Sperelakis N. 1972. (Na^+,K^+)-ATPase activity in tubular systems of mouse cardiac and skeletal muscles. Z Zellforsch Mikroskop Anat 134:1-11.

Fozzard HA. 1972. The membrane capacitance of heart cells. In Electrical Phenomena in the Heart, de Mello WC, eds. New York: Academic, pp 219-233. (Clinical Engineering Series).

Fozzard HA, Kipnis DM. 1967. Regulation of intracellular sodium concentrations in rat diaphragm muscle. Science 156:1257-1260.

Fozzard HA, Lee CO. 1976. Influence of changes in external potassium and chloride ions on membrane potential and intracellular potassium ion activity in rabbit ventricular muscle. J Physiol London 256:663-689.

Frankenheuser B, Hodgkin AL. 1956. The after-effects of impulses in the giant nerve fibres of Loligo. J Physiol London 131:341-376.

Freygang WH Jr, Goldstein DA, Hellam DC, Peachey LD. 1964. The relation between the late after-potential and the size of the transverse tubular system of frog muscle. J Gen Physiol 48:235-263.

Frumento AS. 1965. Sodium pump: its electrical effects in skeletal muscle. Science 147:1442-1443.

Gadsby DC, Kimura J, Noma A. 1985. Voltage dependence of Na/K pump current in isolated heart cells. Nature 315:63-65.

Gadsby DC, Nakao M. 1989. Steady-state current-voltage relationship of the Na/K pump in guinea-pig ventricular myocytes. J Gen Physiol 94:511-537.

Garrahan PJ, Glynn IM. 1967a. Factors affecting the relative magnitudes of the sodium:potassium and sodium:sodium exchanges catalysed by the sodium pump. J Physiol London 192:189-216.

Garrahan PJ, Glynn IM. 1967b. The incorporation of inorganic phosphate into adenosine triphosphate by reversal of the sodium pump. J Physiol London 192:237-256.

Gibbons WR, Fozzard HA. 1971. High potassium and low sodium contractures in sheep cardiac muscle. J Gen Physiol 58:483-510.

Giles W, Shimoni Y. 1989. Comparison of sodium-calcium exchanger and transient inward currents in single cells from rabbit ventricle. J Physiol London 417:465-481.

Giotta GJ. 1976. Quaternary structure of (Na^+,K^+)-dependent adenosine triphosphatase. J Biol Chem 251:1247-1252.

Glitsch HG. 1972. Activation of the electrogenic sodium pump in guinea-pig auricles by internal sodium ions. J Physiol London 220:565-582.

Glitsch HG, Reuter H, Scholz H. 1970. Effect of the internal sodium concentration on calcium fluxes in isolated guinea-pig auricles. J Physiol London 209:25-43.

Glynn IM. 1968. Membrane ATPase and cation transport. Brit Med Bull 24:165-169.

Glynn IM, Hoffman JF, Lew VL. 1971. Some 'partial reactions' of the sodium pump. Phil Trans Roy Soc London Ser B 262:91-102.

Goldin SM. 1977. Active transport of sodium and potassium ions by the sodium and potassium ion-activated adenosine triphosphatase from renal medulla. J Biol Chem 252:5630-5642.

Goldman DE. 1943. Potential, impedance and rectification in membranes. J Gen Physiol 27:37-60.

Goldshlegger R, Karlish SJ, Rephaeli A, Stein WD. 1987. The effect of membrane potential on the mammalian sodium-potassium pump reconstituted into phospholipid vesicles. J Physiol London 387:331-355.

Harris EJ. 1965. The chloride permeability of frog sartorius. J Physiol London 176:123-135.

Harris EJ, Ochs S. 1966. Effects of sodium extrusion and local anesthetics on muscle membrane resistance and potential. J Physiol London 187:5-21.

Harsch M, Green JW. 1963. Electrolyte analyses of chick embryonic fluids and heart tissue. J Cellular Comp Physiol 62:319-326.

Henn FA, Sperelakis N. 1968. Stimulative and protective action of Sr^{2+} and Ba^{2+} on (Na^+,K^+)-ATPase from cultured heart cells. Biochim Biophys Acta 163:415-417.

Herman BA, Fernandez BS. 1976. Developmental changes in membrane fluidity of cultured myogenic cells (Abstract). Physiologist 19:223.

Hermsmeyer K, Rulon R, Sperelakis N. 1972. Loss ofthe plateau of the cardiac action potential in hypertonic solutions. J Gen Physiol 59:779-793.

Hill AV. 1928. The diffusion of oxygen or lactic acid through tissues. Proc Roy Soc London Ser B 104:39-96.

Hiraoka M, Hecht HH. 1973. Recovery from hypothermia in cardiac Purkinje fibers: considerations for an electrogenic mechanism. Pfluegers Arch European J Physiol 339:25-36.

Hodgkin AL. 1964. The Conduction of the Nervous Impulse. Springfield, Ill: Thomas.

Hodgkin AL, Horowicz P. 1959. The influence of potassium and chloride ions on the membrane potential of single muscle fibres. J Physiol London 148:127-160.

Hodgkin AL, Huxley AF. 1939. Action potentials recorded from inside a nerve fibre. Nature 144:710-711.

Hodgkin AL, Huxley AF. 1952a. Currents carried by sodium and potassium ions through the membrane of the giant axon of Loligo. J Physiol London 116:449-472.

Hodgkin AL, Huxley AF. 1952b. A quantitative description of membrane current and its application to conduction and excitation in nerve. J Physiol London 117:500-544.

Hodgkin AL, Katz B. 1949. The effect of sodium ions on the electrical activity of the giant axon of the squid. J Physiol London 108:37-77.

Hodgman CD, Weast RC, Selby SM. 1958. Handbook of chemistry and physics. Chemical Rubber Publishing Co., Cleveland OH.

Hoffman BF, Cranefield PF. 1960. Electrophysiology of the heart. New York: McGraw.

Hokin LE. 1974. Purification and properties of the sodium, potassium ATPase. Ann NY Acad Sci 242:12-23.

Hokin LE, Sastry PS, Galsworthy PR, Yoda A. 1965. Evidence that a phosphorylated intermediate in a brain transport adenosine triphosphatase is an acyl phosphate. Proc Natl Acad Sci US 54:177-184.

Horowicz P, Gerber CJ. 1965. Effects of external potassium and strophanthidin on sodium fluxes in frog striated muscle. J Gen Physiol 48:489-514.

Hutter OF, Noble D. 1961. Anion conductance of cardiac muscle. J Physiol London 157:335-350.

Hutter OF, Warner AE. 1967. The pH sensitivity of the chloride conductance of frog skeletal muscle. J Physiol London 189:403-425.

Huxley AF. 1960. Appendix 2, Compartmental Methods of Kinetic Analysis, by A. K. Solomon. In Mineral Metabolism: An Advanced Treatise, Comar CL, Bronner F, eds. New York: Academic, vol. 1, pt. A, pp 119-167.

Iannaccone ST, Li K-X, Sperelakis N, Lathrop DA. 1989. Insulin-induced hyperpolarization in mammalian skeletal muscle. Am J Physiol 256:C368-C374.

Isenberg G. 1975. Is potassium conductance of cardiac Purkinje fibers controlled by $[Ca^{2+}]_i$? Nature 253:273-274.

Isenberg G, Trautwein W. 1974. The effect of dihydroouabain and lithium ions on the outward current in cardiac Purkinje fibers. Evidence for electrogenicity of active transport. Pfluegers Arch European J Physiol 350:41-54.

Jain MK. 1972. The Bimolecular Lipid Membrane: A System. New York: Van Nostrand.

Jain MK, White FP, Strickholm A, Williams E, Cordes EH. 1972. Studies concerning the possible reconstitution of an active cation pump across an artificial membrane. J Membrane Biol 8:363-388.

January CT, Fozzard HA. 1988. Delayed afterdepolarizations in heart muscle: Mechanisms and relevance. Pharmacol Rev 40:219-227.

Jobsis FF, Vreman HJ. 1973. Inhibition of a Na^+ and K^+ stimulated adenosine triphosphatase by oligomycin. Biochim Biophys Acta 73:346-348.

Jorgensen PL. 1975. Isolation and characterization of the components of the sodium pump. Quart Rev Biophys 7:239-274.

Josephson I, Sperelakis N. 1976. Local anesthetic blockade of inward slow current in cardiac muscle. European J Pharmacol 40:201-208.

Jundt, H, Porzig H, Reuter H, Stucki JW. 1975. The effect of substances releasing intracellular calcium ions on sodium-dependent calcium efflux from guinea-pig auricles. J Physiol London 246:229-253.

Jundt H, Reuter H. 1976. Is sodium-activated calcium efflux from mammalian cardiac muscle dependent on metabolic energy? J Physio. London 266:78P-79P.

Kass RS, Lederer WJ, Tsien RW, Weingart R. 1978. Role of calcium ions in transient inward currents and aftercontractions induced by strophanthidin in cardiac Purkinje fibres. J Physiol London 281:187-208.

Katz B. 1948. The electrical properties of the muscle fiber membrane. Proc Roy Soc London, Ser B 135:506-534.

Katz B. 1966. Nerve, Muscle, and Synapse. New York: McGrawHill.

Katz AM, Louis Cf, Nash-Adler P, Messineo FC, Shigekawa M. 1980. Ca^{2+}-dependent oscillations in the calcium content of cardiac sarcoplasmic reticulum vesicles. Adv Myocardiol 1:173-177.

Kerkut GA, Thomas RC. 1965. An electrogenic sodium pump in snail nerve cells. Comp Biochem Physiol 14:167-183.

Kernan RP. 1962. Membrane potential changes during sodium transport in frog sartorius muscle. Nature 193:986-987.

Kernan RP. 1970. Electrogenic or linked transport. In Membranes and Ion Transport, Bittar EE, eds. New York: Wiley, 1970, vol. 1, pp 395-431.

Keynes RD, Rybova R. 1963. The coupling between sodium and potassium fluxes in frog sartorius muscle. J Physiol London 168:58P.

Kimelberg HK, Papahajopoulos D. 1972. Phospholipid requirements for Na,K-ATPase head-group specificity and fatty acid fluidity. Biochim Biophys Acta 282:277-292.

Kimura J. 1988. Na-Ca exchange and Ca-sensitive non-selective cation current components of transient inward current in isolated cardiac ventricular cells of the guinea pig. J Physiol London 407:79P.

Kimura J, Miyamae S, Noma A. 1987. Identification of sodium-calcium exchange current in single ventricular cells of guinea-pig. J Physiol London 384:199-222.

Kirley TL. 1989. Determination of three disulfide bonds and one free sulfhydryl in the beta subunit of (Na,K)-ATPase. J Biol Chem 264:7185-7192.

Klein RL. 1960. Ontogenesis of K and Na fluxes in embryonic chick heart. Am J Physiol 199:613-618.

Kleyman TR, Cragoe EJ. 1988. Amiloride and its analogs as tools in the study of ion transport. J Membr Biol 105:1-21.

Koch-Weser J. 1963. Influence of osmolarity of perfusate on contractility of mammalian myocardium. Am J Physiol 204:957-962.

Kohlhardt M, Bauer B, Krause H, Fleckenstein A. 1972. Differentiation of the transmembrane Na and Ca channels in mammalian cardiac fibers by use of specific inhibitors. Pfluegers Arch European J Physiol 335:309-322.

Kohlhardt M, Haap K, Figulla HR. 1976. Influence of low extracellular pH upon the Ca inward current and isometric contractile force in mammalian ventricular myocardium. Pfluegers Arch European J Physiol 366:31-38.

Kojima M, Sperelakis N. 1984. Properties of oscillatory afterpotentials in young embryonic chick hearts. Circ Res 55:497-503.

Kostyuk PG, Krishtal OA, Pidoplichko VI. 1972. Potential-dependent membrane current during the active transport of ions in snail neurons. J Physiol London 226:373-392.
Kuffler SW. 1967. Neurolial cells: Physiological properties and a potassium mediated effect of neuronal activity on the glial membrane potential. Proc Roy Soc London, Ser B 168:1-21.
Kushmerick MJ, Podolsky RJ. 1969. Ionic mobility in muscle cells. Science 166:1297-1298.
Kyte J. 1974. The reactions of sodium and potassium ion-activated ATPase with specific antibodies. J Biol Chem 249:3652-3660.
Kyte J. 1975. Structural studies of sodium and potassium ion-activated adenosine triphosphatase. J Biol Chem 250:7443-7449.
Ladle RO, Walker JL. 1975. Intracellular chloride activity in frog heart. J Physiol London 251:549-559.
Lamb JF. 1961. The chloride content of rat auricle. J Physiol London 157:415-425.
Landowne D, Ritchie JM. 1970. The binding of tritiated ouabain to mammalian non-myelinated nerve fibres. J Physiol London 207:529-537.
Langer GA. 1968. Ion fluxes in cardiac excitation and contraction and their relation to myocardial contractility. Physiol Rev 48:708-757.
Lederer WJ, Tsien RW. 1976. Transient inward current underlying arrhythmogenic effects of cardiotonic steroids in Purkinje fibres. J Physiol London 262:73-100.
Lee CO, Fozzard HA. 1975. Activities of potassium and sodium ions in rabbit heart muscle. J Gen Physiol 65:695-708.
Lehmkuhl D, Sperelakis N. 1963. Transmembrane potentials of trypsin-dispersed chick heart cells cultured in vitro. Am J Physiol 205:1213-1220.

Li K-X, Sperelakis N. 1993. Isoproterenol- and insulin-induced hyperpolarization in rat skeletal muscle. J Cell Physiol 157:631-636.

Li K-X, Sperelakis N. 1994. Electrogenic Na-K pump current in rat skeletal myoballs. J Cell Physiol 159:181-186.

Lindenmayer GE, Schwartz A. 1973. Nature of the transport adenosine triphosphatase digitalis complex. J Biol Chem 248:1291-1300.

Lipp P, Pott L. 1988. Transient inward current in guinea-pig atrial myocytes reflects a change of sodium-calcium exchange current. J Physiol London 397:601-630.

Lu H, Lange G, Brooks CMcC. 1965. Factors controlling pacemaker action in cells of the sinoatrial node. Circulation Res 17:460-471.

Luttgau HC, Glitsch HC. 1976. Membrane Physiology of Nerve and Muscle Fibres. Stuttgart: Fischer.

Luttgau HC, Niedergerke R. 1958. Antagonism between Ca and Na ions on the frog's heart. J Physiol London 143:486-505.

Macdonald RL, Mann JE Jr, Sperelakis N. 1974. Derivation of general equations describing tracer diffusion in any two-compartment tissue with application to ionic diffusion in cylindrical muscle bundles. J Theoret Biol 45:107-130.

MacGillivray AD, Hare D. 1969. Applicability of Goldman's constant field assumption to biological systems. J Thero Biol 25:113-126.

MacLennan DH. 1990. Molecular tools to elucidate problems in excitation-contraction coupling. Biophys J 58:1355-1365.

Marban E, Wier WG. 1985. Ryanodine as a tool to determine the contributions of calcium entry and calcium release to the calcium transient and contraction of cardiac Purkinje fibers. Circ Res 56:133-138.

Marmor MF. 1971. The independence of electrogenic sodium transport and membrane potential in a molluscan neurone. J Physiol London 218:599-608.
Martin AR. 1954. The effect of change in length on conduction velocity in muscle. J Physiol Lond 125:215-220.
McAllister RE, Noble D, Tsien RW. 1975. Reconstruction of the electrical activity of cardiac Purkinje fibres. J Physiol London 251:1-59.
Mcdonald TF, MacLeod DP. 1971. Maintenance of resting potential in anoxic guinea pig ventricular muscle: electrogenic sodium pumping. Science 172:570-572.
Mclean MJ, Lapsley RA, Shigenobu K, Murad F, Sperelakis N. 1975. High cyclic AMP levels in young embryonic chick hearts. Develop Biol 42:196-201.
Meech RW. 1974. The sensitivity of Helix aspersa neurones to injected calcium ions. J Physiol London 237:259-277.
Meszaros J, Pappano AJ. 1990. Electrophysiological effects of 1-palmitoylcarnitine in single ventricular myocytes. Am J Physiol 258(Heart Circ Physiology 27):H931-H938.
Moreton RB. 1969. An investigation of the electrogenic sodium pump in snail neurons using the constant-field theory. J Expt Biol 51:181-201.
Mueller P, Rudin DO. 1963. Induced excitability in reconstituted cell membrane structure. J Theoret Biol 4:268-280.
Mullins LJ. 1961. The macromolecular properties of excitable membranes. Ann NY Acad Sci 94:390-404.
Mullins LJ. 1979. The generation of electric currents in cardiac fibers by Na/Ca exchange. Am J Physiol 236(Cell Physiol 5):C103-C110.
Mullins LJ, Brinley FJ. 1969. Potassium fluxes in dialysed squid axons. J Gen Physiol 53:704-740.

Mullins LJ, Noda K. 1963. The influence of sodium-free solutions on the membrane potential of frog muscle fibers. J Gen Physiol 47:117-132.

Nakao M, Gadsby DC. 1986. Voltage dependence of Na translocation by the Na/K pump. Nature 323:628-630.

Nakao M, Gadsby DC. 1989. [Na] and [K] dependence of the Na/K pump current-voltage relationship in guinea-pig ventricular myocytes. J Gen Physiol 94:539-565.

Narahashi T, Moore JW, Scott WR. 1964. Tetrodotoxin blockage of sodium conductance increase in lobster giant axons. J Genysiol 47:965-974.

Nayler WG. 1963. Effect of caffeine on cardiac contractile activity and radiocalcium movement. Am J Physiol 204:969-974.

Nishi S, Soeda H. 1964. Hyperpolarization of a neurone membrane by barium. Nature 204:761-764.

Nishigaki F, Chen T, Hokin LE. 1974. Studies on the characterization of the sodium-potassium transport adenosine triphosphatase. J Biol Chem 249:4911-4916.

Noble D. 1975. The Initiation of the Heartbeat. New York: Oxford Univ. Press.

Noma A, Irisawa H. 1974. Electrogenic sodium pump in rabbit sinoatrial node cell. Pfluegers Arch European J Physiol 351:177-182.

Pagano R, Thompson TE. 1968. Spherical lipid bilayer membranes: electrical and isotopic studies of ion permeability. J Mol Biol 38:41-57.

Page E, Storm SR. 1965. Cat heart muscle in vitro. VIII. Active transport of sodium in papillary muscles. J Gen Physiol 48:957-972.

Pappano AJ. 1972. Sodium-dependent depolarization of non-innervated embryonic chick heart by acetylcholine. J Pharmacol Exptl Therap 180:340-350.

Pappano AJ, Sperelakis N. 1969. Low K^+ conductance and low resting potentials of isolated single cultured heart cells. Am J Physiol 217:1076-1082.

Pappano AJ, Tanaka M, Meszaros J. 1990. Palmitoylcarnitine, a metabolite of ischemia, inhibits sodium pump current in guinea pig ventricular myocytes. In Ionic currents and ischemia. Vereecke J, vanBogaert PP, Verdonck F, eds. Leuven: University Press, 1990, pp. 205-207.

Pedemonte CH, Kaplan JH. 1990. Chemical modification as an approach to elucidation of sodium pump structure-function relations. Am J Physiol 258:C1-C23.

Pelleg A, Vogel S, Belardinelli L, Sperelakis N. 1980. Overdrive suppression of automaticity in cultured chick myocardial cells. Am J Physiol/Heart & Circulatory Physiol 238:H24-H30.

Philipson KD. 1984. Interaction of charged amphiphiles with Na^+-Ca^{2+} exchange in cardiac sarcolemmal vesicles. J Biol Chem 259:13999-14002.

Philipson KD, Reeves JP. 1992. Cloning and expression of the bovine cardiac sodium-calcium exchanger. Biochem Biophys 298:553-560.

Pitts BJR, Tate CA, VanWinkle B, Wood JM, Entman ML. 1978. Palmitylcarnitine inhibition of the calcium pump in cardiac sarcoplasmic reticulum: A possible role in myocardial ischemia. Lif Sci 23:391-402.

Post RL, Merritt CR, Kinsolving CR, Albright CD. 1960. Membrane adenosine triphosphatase as a participant in the active transport of sodium and potassium in the human erythrocyte. J Biol Chem 235:1796-1802.

Post RL, Sen AK, Rosenthal AS. 1965. A phosphorylated intermediate in the ATP-dependent sodium and potassium transport across kidney membranes. J. Biol. Chem. 240:1437-1445.

Prosser CL, Weems WA, Connor JA. 1976. Types of slow rhythmic activity in gastrointestinal muscles. In

Physiology of Smooth Muscle, Bulbring E, Shuba MF, eds. New York: Raven, pp 99-109.

Rakowski RF, Gadsby DC, DeWeer P. 1989. Stoichiometry and voltage dependence of the sodium pump in voltage clamped, internally dialysed squid giant axon. J Gen Physiol 93:903-841.

Rakowski RF, Paxson CL. 1988. Voltage dependence of Na/K pump current in Xenopus oocytes. J Membr Biol 106:173-182.

Rang HP, Ritchie JM. 1968. On the electrogenic sodium pump in mammalian non-myelinated nerve fibres and its activation by various external cations. J Physiol London 196:183-221.

Rapoport SI. 1970. The sodium-potassium exchange pump: relation of metabolism to electrical properties of the cell. I Theory Biophys J 10:246-259.

Reporter M. 1972. An ATP pool with rapid turnover within the cell membrane. Biochem Biophys Res Commun 48:598-604.

Reuben JP, Girardier L, Grundfest H. 1962. The chloride permeability of crayfish muscle fibres. Biol Bull 123:509-510.

Reuter H. 1974. Exchange of calcium ions in the mammalian myocardium. Circulation Res 34:599-605.

Reuter H, Seitz N. 1968. Dependence of calcium efflux from cardiac muscle on temperature and external ion composition. J Physiol London 195:451-470.

Ritchie JM, Straub RW. 1957. The hyperpolarization which follows activity in mammalian nonmedullated fibres. J Physiol London 136:80-97.

Robinson RA, Stokes RH. 1959. Electrolyte solutions. London: Buterworth & Co.(vol. 2).

Romanoff A. 1960. The Avian Embryo, Structure and Functional Development. New York: Macmillan.

Rougier OG. Vassort G, Garnier D, Gargouil YM, Coraboeuf E. 1969. Existence and role of a slow inward current during the frog atrial action potential. Pfluegers Arch European J Physiol 308:91-110.

Sakata K, Hayashi H, Kobayashi A, Yamazaki N. 1989. Mechanism of arrhythmias induced by palmitylcarnitine in guinea pig pappillary muscle. Cardiovasc Pharmacol 23:505-511.

Schneider JA, Sperelakis N. 1974. The demonstration of energy dependence of the isoproterenol-induced transcellular Ca^{2+} current in isolated perfumed guinea pig hearts -an explanation for mechanical failure of ischemic myocardium. J Surg Res 16:389-403.

Schneider JA, Sperelakis N. 1975. Slow Ca^{2+} and Na^{+} responses induced by isoproterenol and methylxanthines in isolated perfused guinea pig hearts exposed to elevated K^{+}. J Mol Cellular Cardiol 7:249-273.

Schultze W, Wollengerger A. 1969. Zytochemische Localisation und Differenzierung von Na^{+},K^{+}-und anderer membranstandiger ATPase-aktivitat im Herzmuskel. Histochemie 19:302-318.

Schwartz A. 1962. A sodium and potassium stimulated ATPase from cardiac tissue. I. Preparation and properties. Biochem Biophys Res Commun 9:301-306.

Schwartz A, Lindenmayer GE, Allen JC. 1975. The sodium-potassium adenosine triphosphatase: pharmacological, physiological and biochemical aspects. Pharmacol Rev 27:3-134.

Schweigert B, Lafaire AV, Schwarz W. 1988. Voltage dependence of the Na-K ATPase: Measurements of ouabain-dependent membrane current and ouabain binding in oocytes of Xenopus laevis. Pflugers Arch 412:579-588.

Sellin LC, Sperelakis N. 1978. Decreased potassium permeability in dystrophic mouse skeletal muscle. Exptl Neurology 62:605-617.

Sen AK, Post RL. 1964. Stoichiometry and localization of adenDsine triphosphate-dependent sodium and potassium transport in erythrocyte. J Biol Chem 239: 345-352.

Sha'afi RI, Rodan SB, Hintz RL. Fernandez SM, Rodan GA. 1975. Abnormalities in microviscosity and ion transport in genetic muscular dystrophy. Nature 254: 525-526.

Shigenobu K, Schneider JA, Sperelakis N. 1974. Blockade of slow Na^+ and Ca^{++} currents in myocardial cells by verapamil. J. Pharmacol. Exptl. Therap. 190:280-288.

Shigenobu K, Sperelakis N. 1972. Ca^{++} current channels induced by catecholamines in chick embryonic hearts whose fast Na^+ channels are blocked by tetrodotoxin or elevated K^+. Circulation Res. 31: 932-952.

Shimizu Y, Tasaki K. 1965. Electrical excitability of developing cardiac muscle in chick embryos. Tohoku J Exptl Med 88:49-56.

Singer SJ, Nicolson GL, 1972. The fluid mosaic model of the structure of cell membranes. Science 175:720-731.

Sjodin RA, Beauge LA, 1968. Coupling and selectivity of sodium and potassium transport in squid giant axons. J Gen Physiol 51:152S-161S.

Sjodin RA, Ortiz O. 1975. Resolution of the potassium ion pump in muscle fibers using barium ions. J Gen Physiol 66:269-286.

Skou JC. 1957. The influence of some cations on an adenosine triphosphatase from peripheral nerves. Biochim Biophys Acta 23:394-401.

Skou JC. 1965. Enzymatic basis for active transport of Na^+ and K^+ across cell membrane. Physiol Rev 45:596-617.

Skou JC. 1971. The role of the phosphorylated intermediates in the reaction of the Na^+,K^+-activated enzyme system. In Membrane-Bound Enzymes, Porcellati G, DiJeso F, eds.

New York: Plenum, pp 175-195 (Advances in Experimental Medicine and Biology Series, vol. 14).

Skou JC. 1974. The role of ATP in the reaction of Na^+,K^+-ATPase system. Ann NY Acad Sci 242:168-184.

Skou JC. 1975. The (Na+,K^+) activated enzyme system and its relationship to transport of sodium and potassium. Quart Rev Biophys 7:401-434.

Skou JC, Zerahn K. 1959. Investigations on the effect of some local anesthetics and other amines on the active transport of sodium through the isolated short-circuited frog skin. Biochim Biophys Acta 35:324-333.

Sperelakis N. 1969. Changes in conductances of frog sartorius fibers produced by $C0_2$, ReO_4, and temperature. Am J Physiol 217:1069-1075.

Sperelakis N. 1972a. The electrical properties of embryonic heart cells. In *Electrical Phenomena in the Heart,* de Mello WC, ed. New York: Academic, pp 1-61.

Sperelakis N. 1972b. (Na^+,K^+)-ATPase activity of embryonic chick heart and skeletal muscles as a function of age. Biochim Biophys Acta 266:230-237.

Sperelakis, N. 1979. Origin of the cardiac resting potential. In: *Handbook of Physiology, The Cardiovascular System,* Vol. 1: The Heart (R.M. Berne and N. Sperelakis, eds.), Am. Physiol. Soc., Chap. 6, pp. 187-267, .

Sperelakis N. 1985. Pathophysiology of skeletal muscle and effect of some hormones. In The Thorax: Vital Pump, Roussos Ch, Macklem P, eds. Marcell Dekker, Inc., pp 115-140.

Sperelakis, N. 1993. Cable properties and propagation. Chap. 5, pp. 83-97, In Essentials of Physiology, (N. Sperelakis & R.O. Banks, eds.); Little, Brown, and Co., Boston.

Sperelakis N, Henn FA. 1970. Effect of diphenylhydantoin on membrane potentials and Na^+,K^+-ATPase of cultured chick heart cells. Am J Physiol 218:1224-1227.

Sperelakis N, Lee, EC. 1971. Characterization of (Na^+,K^+)-ATPase isolated from embryonic chick hearts and cultured chick heart cells. Biochim Biophys Acta 233:562-579.

Sperelakis N, Lehmkuhl D. 1966. Ionic interconversion of pacemaker and nonpacemaker cultured chick heart cells. J Gen Physiol 49:867-895.

Sperelakis N, LoBrocco B, Mann JE, Marschall R. 1985. Potassium accumulation in intercellular junctions combined with electric field interactions for propagation in cardiac muscle. Innov et Tech en Biol et Med 6(1):24-43.

Sperelakis N, Mayer G, Macdonald R. 1970. Velocity of propagation in vertebrate cardiac muscles as functions of tonicity and $[K^+]_o$. Am J Physiol 219:952-963.

Sperelakis N, McLean MJ. 1978a. Electrical properties of cultured heart cells. In Recent Advances in Studies on Cardiac Structure and Metabolism, Kobayashi T, Ito Y, Rona G, eds. Baltimore: Univ. Park, vol. 12, pp 645-660.

Sperelakis N, McLean MJ. 1978b. The electrical properties of embryonic chick cardiac cells. In Fetal and Newborn Cardiovascular Physiology, Longo LD, ed. New York: Garland, vol. 1, pp 191-236.

Sperelakis N, Rubio R. 1971. Ultrastructural changes produced by hypertonicity in cat cardiac muscle. J Mol Cellular Cardiol 3:139-156.

Sperelakis N, Schneider JA. 1976. A metabolic control mechanism for Ca^{++} influx which may protect the myocardial cell. Am J Cardiol 37:1079-1085.

Sperelakis N, Schneider MF. 1968. Membrane ion conductances of frog sartorius fibers as a function of tonicity. Am J Physiol 215:723-729.

Sperelakis N, Schneider MF, Harris EJ. 1967. Decreased K^+ conductance produced by Ba^{++} in frog sartorius fibers. J Gen Physiol 50:1565-1583.

Sperelakis N, Shigenobu K. 1972. Changes in membrane properties of chick embryonic hearts during development. J Gen Physiol 60:430-453.

Sperelakis N, Shigenobu K, Mclean MJ. 1975. Membrane cation channels -Changes in developing hearts, in cell culture, and in organ culture. In Developmental and Physiological Correlates of Cardiac Cells, Lieberman M, Sano T, eds. New York: Raven, 1975, pp 209-234.

Sperelakis N, Tarr M. 1965. Weak electrotonic interaction between neighboring visceral smooth muscle cells. Am J Physiol 208:737-747.

Sperelakis N, Valle R, Orozco C, Martinez-Palomo A. Rubio R. 1973. Electromechanical uncoupling of frog skeletal muscle by possible change in sarcoplasmic reticulum content. Am J Physiol 225:793-800.

Tahara T, Kimizuka H, Koketsu K. 1973. An analysis of the membrane hyperpolarization during action of the sodium pump in frog's skeletal muscles. Japan J Physiol 23:165-181.

Tanaka M, Gruver CL, Pappano AJ. 1989. 1-palmitoylcarnitine inhibits Na/K pump in isolated ventricular myocytes. Circulation 80:II-304.

Thakkar JK, Sperelakis N. 1987. Changes in cyclic nucleotide levels during embryonic development of chick hearts. J Devel Physiol 9:497-505.

Thomas RC. 1969. Membrane current and intracellular sodium changes in a snail neurone during extrusion of injected sodium. J Physiol London 201:495-514.

Thomas RC. 1972. Electrogenic sodium pump in nerve and muscle cells. Physiol Rev 52:563-594.

Thompson TE, Henn FA. 1970. Experimental phospholipid model membranes. In Membranes of Mitochondria and Chloroplasts, Racker E, ed. New York: Van Nostrand, pp 1-52.

Thureson-Klein A, Klein RL. 1971. Cation distribution and cardiac jelly in early embryonic heart: a histochemical and electron microscopic study. J Mol Cellular Cardiol 2:31-40.

Tsien RW, Giles W, Greengard P. 1972. Cyclic AMP mediates the effects of adrenaline on cardiac Purkinje fibres. Nature New Biol 240:181-183.

Ussing HH. 1949. The distinction by means of tracers between active transport and diffusion. Acta Physiol Scand 19:43-56.

Van Liew HD. 1970. Graphic analysis of aggregates of linear and exponential processes. J Theoret Biol 16:43-53.

Vassalle M. 1970. Electrogenic suppression of automaticity in sheep and dog Purkinje fibers. Circulation Res 27:361-377.

Vaughan-Jones RD. 1977. Intracellular chloride activity of quiescent cardiac Purkinje fibres. J Physiol London 272:32P-33P.

Vogel S, Sperelakis N. 1977. Blockade of myocardial slow inward current at low pH. Am J Physiol 233:C99-ClO3.

Walker JL, Brown HM. 1977. Intracellular ionic activity measurements in nerve and muscle. Physiol Rev 57:729-778.

Weidmann S. 1956. Elektrophysiologie der herzmuskelfaser. Bern: Huber.

Wiggins JR, Cranefield PF. 1976. Two levels of resting potential in canine cardiac Purkinje fibers exposed to sodium-free solutions. Circulation Res 39:466-474.

Wilbrandt W, Koller H. 1948. Die Calcium-Wirkung am Froschherzen als Funktion des Ionengleichgewichts zwischen Zellmembran und Umgebung. Helv Physiol Pharmacol Acta 6:208-221.

Wood JM, Bush B, Pitts BJR, Schwartz A. 1977. Inhibition of bovine heart Na^{+}-K^{+}-ATPase by palmitylcarnitine and palmityl-CoA. Biochem Biophys Res Commun 74:677-684.

Yeh BK, Hoffmann BK. 1967. The ionic basis of electrical activity in embryonic cardiac muscle. J Gen Physiol 52:666-681.

INDEX

GPSR Compliance

The European Union's (EU) General Product Safety Regulation (GPSR) is a set of rules that requires consumer products to be safe and our obligations to ensure this.

If you have any concerns about our products, you can contact us on ProductSafety@springernature.com

In case Publisher is established outside the EU, the EU authorized representative is:

Springer Nature Customer Service Center GmbH
Europaplatz 3
69115 Heidelberg, Germany

Batch number: 10755284

Printed by Printforce, the Netherlands